T0329841

Optimal Transport Methods in Economics

Optimal Transport
Methods in Economics

Alfred Galichon

PRINCETON UNIVERSITY PRESS

PRINCETON AND OXFORD

Published by Princeton University Press
41 William Street, Princeton, New Jersey 08540

In the United Kingdom: Princeton University Press
6 Oxford Street, Woodstock, Oxfordshire, OX20 1TR

Jacket art by Elijah Meeks, Stanford University Library

ISBN: 978-0-691-17276-7
Library of Congress Control Number: 2016933358

British Library Cataloging-in-Publication Data is available

This book has been composed in Linux Libertine

press.princeton.edu

Typeset by Nova Techset Pvt Ltd, Bangalore, India

1   3   5   7   9   10   8   6   4   2

To Audrey, Jacqueline, and André

Contents

Preface

I started working on these lecture notes for a graduate course I gave at MIT in the Spring 2015 semester. They provide an introduction to the theory of optimal transport, with a focus on applications to economic modeling and econometrics. They are intended to cover basic results in optimal transport, in connection with linear programming, network flow problems, convex analysis, and computational geometry. Several applications to various fields in economic analysis (econometrics, family economics, labor economics, and contract theory) are provided.

Optimal transport and its applications to economics and statistics have been at the center of my own research in the last ten years. I discovered the topic with Cédric Villani's fascinating textbook, *Topics in Optimal Transportation*, shortly after it appeared, and it has been an inspiration for the present project. I was lucky enough to deepen my knowledge of the subject in a series of lectures taught by Ivar Ekeland as part of workshops at Columbia University and the University of British Columbia; a wonderful correspondence and collaboration followed. My intellectual debt to these two masters is huge. Over the years, I have been lucky enough to work on both theoretical and applied aspects of optimal transport with talented researchers, from whom I have learned a lot and to whom I extend my warm thanks. They are (in alphabetical order) Raicho Bojilov, Odran Bonnet, Damien Bosc, Guillaume Carlier, Arthur Charpentier, Victor Chernozhukov, Pierre-André Chiappori, Khai Chiong, Edoardo Ciscato, Rose-Anne Dana, Arnaud Dupuy, Federico Echenique, Ivar Ekeland, Ivan Fernandez-Val, Denis Fougère, Nassif Ghoussoub, Marion Goussé, Marc Hallin, Marc Henry, Pierre Henry-Labordère, Yu-Wei Hsieh, Sonia Jaffe, Scott Kominers, Alex Kushnir, Mathilde Poulhès, Bernard Salanié, Filippo Santambrogio, Matt Shum, Nizar Touzi, Simon Weber, and Liping Zhao. Besides them, I also have enjoyed enlightening conversations with Robert McCann. I would like to thank Sarah Caro, my editor at Princeton University Press, for her constructive support, and her editorial team, including Hannah Paul, her assistant, and Alison Durham, the copyeditor, for their help with the publication process. I am also grateful to Pauline Corblet, Arthur Morisseau Duprat de Mézailles, Keith O'Hara, and Simon Weber for their extensive reading of the manuscript and valuable comments.

My research has received funding from the European Research Council under the European Union's Seventh Framework Programme (FP7/2007-2013)/ ERC grant agreements no. 313699. Most of this book was written while I was on leave at MIT in 2014–2015. I thank the Economics Department there for their hospitality.

Optimal Transport Methods in Economics

—1—

Introduction

The basic problem in optimal transport (hereafter, OT) can be best exemplified by the problem of assigning workers to jobs: given the distribution of a population of workers with heterogenous skills, and given the distribution of jobs with heterogeneous characteristics, how should one assign workers to firms in order to maximize the total economic output? The economic output, of course, will depend on the complementarities between workers' skills and job characteristics; some assignments generate higher total output than others. This problem and its variants are known under several names: mass transportation, optimal assignment, matching with transferable utility, optimal coupling, Monge–Kantorovich, and Hitchcock being the more common. Of course, the multiplicity of names reflects small variations in the formulation of the problem, but also the stunning diversity of applications this theory has found.

1.1 A NUMBER OF ECONOMIC APPLICATIONS

To describe OT as a general framework for labor market assignment problems, as we just did, is somewhat overrestrictive. While labor economics is certainly one use of the theory, which we will discuss below, there is much more to it. Indeed, an impressive number of seemingly unrelated problems in economics have the structure of an OT problem. Here are some examples, without any attempt at exhaustivity.

  – *Matching models* are models in which two populations, such as men and women on the marriage market, workers and machines etc., must be assigned into pairs. Each pair generates a surplus which depends on the characteristics of both partners. One question deals with the characterization of the *optimal assignment*: what is the assignment a central planner would choose in order to maximize the total utility surplus? Another question deals with the *equilibrium assignment*: letting partners match in a decentralized manner, what are the equilibrium matching patterns and transfers? The Monge–Kantorovich theorem, introduced in chapter 2, implies that the answers to these two questions coincide: any optimal solution chosen by the central planner can be obtained at equilibrium and conversely, any equilibrium assignment is also optimal. Thus OT theory will provide a powerful welfare theorem in matching models with transferable utility.

– *Models of differentiated demand* are models where consumers who choose a differentiated good (say, a house) have unobserved variations in preferences. These types of models are often called *hedonic models* when the measure of the quality of the good is continuous, and *discrete choice models* when it is discrete. A central problem in these models is the identification of preferences. By imposing assumptions on the distribution of the variation in preferences, one is able to identify the preferences on the basis of distribution of the demanded qualities. It turns out that these preferences happen to be the solution to the dual Monge–Kantorovich problem. This approach is explained in sections 9.2 and 9.4. In this context, OT therefore provides a constructive identification strategy.

– Some *incomplete econometric models* can be addressed using OT theory. In some problems, data are incomplete or missing, which creates a partial identification issue. For instance, income is sometimes reported only in tax brackets, therefore a model using the distribution of income as a source of identification may be incomplete in the sense that several values of the parameters may be compatible with the observed distribution. The problem of determining the identified set, namely, the set of parameters compatible with the observed distribution, can be reformulated as an OT problem. OT problems enjoy nice computational properties that make them efficiently computable. Hence the OT approach to partial identification is practical, as it allows fast computation of the identified set. Section 9.1 elaborates on this.

– *Quantile methods* are useful econometric and statistical techniques for analyzing distributions and dependence between random variables. They include among others quantile regression, quantile treatment effect, and least absolute deviation estimation. In dimension one, a quantile map is simply the inverse of the cumulative distribution function. As we shall see in chapter 4, quantile maps are very closely connected to an OT problem. In particular, OT provides a way to define a generalization of the notion of a quantile; see sections 9.4 and 9.5.

– In *contract theory*, multidimensional principal–agents problems may be reformulated as OT problems, as seen in section 9.6. This reformulation has useful econometric implications, as it allows us to infer each agent's unobserved type based on the observed choices, assuming that the distribution of types is known.

– Some *derivative pricing* questions can be answered using OT, in particular the problem of bounds on derivative prices. A derivative is a financial asset whose value depends on the value of one or several other traded assets, called underlyings. Derivatives with several underlyings are often hard to price in practice, as their value depends not only on the distribution of the value of each underlying, but also on the joint distribution. Often, the distribution of each underlying is known, as it can be recovered from market prices. The Monge–Kantorovich theorem is then useful to analyze bounds of prices of derivatives with multiple underlyings. An example of this method is provided in section 9.7. Similarly, OT is useful in *risk management* problems, where the measure of a given risk often depends on the joint distribution of several risks whose marginal distribution is known but whose dependence structure is unknown. Providing bounds on these measures of risk can then be rephrased as an OT problem.

1.2 A MIX OF TECHNIQUES

Surprisingly, in several other scientific disciplines, OT has allowed old problems to be revisited, and has brought new insights into them. This is the case in astrophysics (where OT has been used to model the early universe), in meteorology (where it has been used to model atmospheric fronts), in image analysis (where it provides convenient interpolation tools), and even in pure mathematics (it is insightful for analyzing Ricci curvature in Riemannian geometry). What is the reason for this apparent universality? Why is OT so prevalent? One answer may come from the very strong link between OT and convex analysis. Convex analysis is a most useful tool in many sciences, and OT is a way to revisit convex analysis in depth. As we shall see, one can learn about convex analysis almost entirely from the sole point of view of OT. In fact, the latter allows—at no extra cost—a significant generalization of convex analysis, which will be described in section 7.1. Hence, it should not be surprising that many problems in economics and other disciplines have a natural reformulation as an OT problem.

Moreover, developing an in-depth knowledge of OT will help the reader to discover, or rediscover, a number of tools. Indeed, one interesting feature of OT, especially from the point of view of a student eager to learn useful techniques "on the go," is that it is connected to a number of important methods from various fields. OT is a mix of different techniques, and this text will contain a number of "crash courses" on a variety of topics. Let us briefly discuss a few of these topics and how they will occur in the book.

– *Linear programming* will underlie much of these notes. While optimal transportation in the general (continuous) case is an infinite-dimensional linear programming problem, and needs to be studied with more specific tools, we will see in chapter 3 that it boils down in the discrete case to a prototypical linear programming problem. In chapter 3 and appendix B we will spell out the basics of linear programming, with no prerequisite knowledge on the topic.

– More generally, this book will make heavy use of *large-scale optimization* methods. Indeed, the optimal transportation problem is a linear programming problem of a particular sort in the sense that it has a very sparse structure: the matrix of constraints contains many zeros. When computing these problems using linear programming algorithms, this fact calls for the use of large-scale optimization techniques, which take the sparsity of the matrix of constraints into consideration. We will demonstrate the interest of recognizing the sparse structure of the problem by giving computational examples written in R interfaced with Gurobi, a state-of-the-art linear programming solver.

– *Convex analysis* will be met in several places within this book. In the first place, OT problems are convex optimization problems, as are all linear programming problems. Also, we will see in section 6 that a special case of the OT problem, when the surplus is quadratic and the distributions are continuous, yields solutions that are convex functions. Chapter 6 will then be the occasion to revisit convex analysis from the point of view of optimal transport.

– The general setting of *network flow problems* will be studied in chapter 8. These extend discrete OT problems, and are among the most useful and best-studied problems in operations research. In a minimum cost flow problem, one seeks to send mass from a number of source nodes to a number of demand nodes through the network along paths of intermediate nodes in a way which minimizes the total transportation cost. Minimum cost flow problems combine a shortest path problem (find the cheapest path from one supply to one demand node) and an OT problem (find the optimal assignment between supply and demand nodes associated with the optimal cost between any pair of nodes). There is a continuous extension of this theory—not discussed in these notes—whose cornerstone result is McCann's theorem on optimal transportation on manifolds.

– Finally, these notes will also incidentally feature some tools for *spatial data analysis* and *computational geometry* (introducing Voronoi cells, power diagrams, and the Hotelling location game); for *supermodularity*; and for *matrix theory*. The list goes on, but it should by no means discourage the reader. Again, these notes were written to be as self-contained as possible, in the hope that the reader will develop a working knowledge of the mix of techniques that is required for an in-depth understanding of OT.

1.3 BRIEF HISTORY

The history of OT starts with a French mathematician and statesman, Gaspard Monge (1746–1818, see figure 2.1), who is also the inventor of descriptive geometry, and the founder of École Polytechnique. Monge formulated the problem for the first time in 1781 out of civil engineering preoccupations. As we will see in chapter 2, Monge was concerned with a particularly difficult variant of the problem, and the solution he gave was incomplete. Despite significant efforts, nineteenth-century mathematicians failed to overcome the difficulty. The problem remained unsolved until 1941, when the great Soviet mathematician Leonid Kantorovich (1912–1986, see figure 2.2), and independently a few years after him, Koopmans and his collaborators, introduced the relaxation technique described in chapter 2, allowing the problem to be relaxed into a linear programming problem. Duality provided a powerful tool to analyze the problem and its properties, and to provide an economic interpretation. The second half of the twentieth century mostly focused on the discrete assignment problem, detailed in chapter 3. It was only at the end of the 1980s and in the 1990s, with the work of Brenier, Knott, Rachev, Rüschendorf, Smith, McCann, Gangbo, and others, that Kantorovich duality was put to efficient use to fully solve Monge's problem with quadratic costs and make a complete connection with convex duality, as described in chapter 6. This discovery, whose most striking formulation is Brenier's theorem, presented in section 6.2, sparked a renewed interest in the topic, and OT is currently a very active area of research in mathematics and many applied sciences.

The interest in the numerical computation of OT problems (in its finite-dimensional version) is almost as old as the problem itself, although it was studied independently. It seems that the first efficient assignment algorithm (known today as the "Hungarian algorithm") was discovered by Carl Jacobi around 1850, and later rediscovered in the 1950s with the work of Kuhn, Munkres, Kőnig, and Egerváry.

1.4 LITERATURE

There are good sources on OT in the mathematical literature. Primary references include two excellent, and fairly recent monographs by Cédric Villani. *Topics in Optimal Transportation* [148] is a set of great lecture notes, written in an intuitive way. The intellectual debt the present text owes to the former cannot be overstated. We will very often refer to it, and suggest that the reader should study both texts in parallel. *Optimal Transport: Old and New* [149] is the definitive reference on the topic, written in a more encyclopedic way, which is why we do not recommend it as an introduction. Both of these books, even the first one, require a significant investment; further, the author's favorite application is fluid mechanics rather than economics. For these reasons, economists do not always find it easy to appropriate this material. Another good source is Rachev and Rüschendorf's two volume treatise *Mass Transportation Problems* [118]. Although somewhat outdated given the progress in the literature over the last twenty years, it contains insightful discussions and examples that are not found in Villani's texts. Finally, Santambrogio's recent book, *Optimal Transport for Applied Mathematicians* [132], has an original perspective on the topic and offers some interesting computational considerations. None of these texts has a focus on economics; in contrast, at least two high quality sets of introductory lecture notes are explicitly aimed at economic applications, despite being written in the mathematical tradition. One is Carlier's unpublished 2010 lecture notes [32], which can easily be found online. The other one is Ekeland's lecture notes [50], which appeared in 2010 in a special issue of *Economic Theory* dedicated to economic applications of OT.

On the other hand, the economic literature is somewhat terse on up-to-date reference texts for OT and its economic applications. There is little or no mention of the topic in the main graduate microeconomics textbooks. The classic treatise by Roth and Sotomayor [125] on two-sided matching deals mostly with models with nontransferable utility, that do not belong in the category of OT problems. It has one excellent chapter on the optimal assignment in the finite-dimensional (i.e., discrete) case, with which chapter 3 of the present text partially overlaps, but it covers none of the more advanced topics. Vohra [150] has excellent coverage of much of the basic mathematical machinery needed for the present book, and has a concise, yet informative section on assignment problems, but is also restricted to the finite-dimensional case. Another enlightening text by the same author, Vohra [151], offers a unifying reformulation of mechanism design theory using network flows; however, it also predominantly focuses on the discrete case, and it deals only with mechanism design applications.

Given the central importance of OT in the field, it is somewhat strange that the economic literature is missing an introductory text on the topic. We believe that the time has come for such a book.

1.5 ABOUT THESE NOTES

The purpose of the present notes is precisely to guide economists through this topic, by highlighting the potential for economic applications, and cutting short the part of the theory which is not of primordial importance for the latter. These notes are therefore intended as a complement to, rather than as a substitute for Villani's text mentioned above, [148], which we strongly suggest that the reader should read in parallel.

Because the purpose of this book is not to replace [148], but rather to complement it, this book is written in such a way that the formal statements, theorems and propositions, will be mathematically correct, but the proofs will sometimes be only sketched, or sometimes be shown under a set of stronger assumptions. Our style of exposition therefore draws inspiration from texts on mathematical physics, and we will at times content ourselves with a "sketch of proof" explaining why a result is true without providing a proof that mathematicians would receive as acceptable. As a result, this text will be self-contained for readers who are content with results and their economic intuition; but readers who want to see full proofs will often be referred to [148], or Villani's other monograph [149].

Another contrast between Villani's text and the present one is the focus on computation in the latter. Economists (or, more precisely, econometricians) need to take their models to data. Economists are happy to know about the existence of a solution, but they worry if they cannot compute it in a reasonable amount of time. Complementing mathematical results with algorithms is quite natural as OT problems are closely linked to linear programming and optimal assignments, which are computationally tractable optimization problems for which there are well-developed efficient solutions. Hence, computation will be inherently part of this book, and examples labeled "Programming Example" will provide details on implementations. Our approach, however, has been strongly biased. Rather than looking for the most efficient method adapted to a given particular problem, we have sought to demonstrate that general purpose linear programming techniques, combined with the use of libraries to handle sparse matrices and matrix algebra, yield satisfactory results for most applications we will discuss. The demonstration codes are therefore written in R, which is an open-access mathematical programming language and allows easy and quick prototyping of programs. Although it is not advisable to use R directly for optimization, it can easily be interfaced with most optimization solvers. We will make frequent use of Gurobi, a state-of-the- art linear programming solver, which is commercial software, but is provided for free to the academic community (www.gurobi.com). The full set of programs is provided via this book's web page at http://press.princeton.edu/titles/10870.html. The reader is strongly encourage to try their own programs and compare them with those provided.

As this text is intended as a graduate course, and as learning requires practice, a number of exercises are provided throughout the book. Some of them (labeled "M") are intended to develop mathematical agility; some (labeled "C") help the reader to get used to computational techniques; others (labeled "E") are intended to build economic intuition.

1.6 ORGANIZATION OF THIS BOOK

Let us briefly summarize the content of each chapter.

Chapter 2 states the Monge–Kantorovich problem and provides the duality result in a fairly general setting. The primal problem is interpreted as the central planner's problem of determining the optimal assignment of workers to firms, while the dual problem is interpreted as the invisible hand's problem of obtaining a system of decentralized equilibrium prices. In general, the primal problem always has a solution (which means that an optimal assignment of workers to jobs exists), but the dual does not: the optimal assignment cannot always be decentralized by a system of prices. However, as we shall see later, the cases where the dual problem does not have a solution are rather pathological, and in all of the cases considered in the rest of the book, both the primal and the dual problems have solutions.

Chapter 3 considers the finite-dimensional case, which is the case when the marginal probability distributions are discrete with finite support. In this case the Monge–Kantorovich problem becomes a finite-dimensional linear programming problem; the primal and the dual solutions are related by complementary slackness, which is interpreted in terms of stability. The solutions can be conveniently computed by linear programming solvers, and we will show how to do this using some matrix algebra and Gurobi.

Chapter 4 considers the univariate case, when both the worker and the job are characterized by a scalar attribute. The important assumption of positive assortative matching, or supermodularity of the matching surplus, will be introduced and discussed. As a consequence, the primal problem has an explicit solution (an optimal assignment) which is related to the probabilistic notion of a quantile transform, and the dual problem also has an explicit solution (a set of equilibrium prices), which are obtained from the solution to the primal problem. As a consequence, the Monge–Kantorovich problem is explicitly solved in dimension one under the assumption of positive assortative matching.

Chapter 5 considers the case when the attributes are d-dimensional vectors, the matching surplus is the scalar product, the distribution of workers' attributes is continuous, and the distribution of the firms is discrete. The geometry of the optimal assignment of workers to firms is discussed and related to the important notion of power diagrams in computational geometry. The optimal assignment map is shown to be the gradient of a piecewise affine convex function, and the equilibrium prices of the firms are shown to be the solution to a finite-dimensional convex minimization problem. We will discuss how to perform this computation in practice.

Chapter 6 still considers the case when the attributes are d-dimensional vectors and the surplus is the scalar product; it still assumes that the distribution of the workers' attributes is continuous, but it relaxes the assumption that the distribution of the firms' attributes is discrete. This setting allows us to entirely rediscover convex analysis, which is introduced from the point of view of optimal transport. As a consequence, Brenier's polar factorization theorem is given, which provides a vector extension for the scalar notions of quantile and rearrangement.

Chapter 7 considers a case with a more general surplus function. This is the place to show that when the scalar-product surplus is replaced by a more general function, much of the machinery put in place in chapter 6 goes through. In particular, it is possible to generalize convex analysis in a natural way, and to obtain generalized notions of convex conjugates, of convexity, and of a subdifferential that are perfectly suited to the problem. A general result on the existence of dual minimizers will be given, as well as sufficient conditions for the existence of a solution to the Monge problem.

Chapter 8 considers the optimal network flow problem, which is a generalization of the optimal assignment problem considered in chapter 3. In optimal flow problems, one considers a network of cities, or edges, to move a distribution of mass on supply nodes to a distribution of mass on demand nodes. The difference from a standard optimal assignment problem is that the matching surplus associated with moving from a supply location to a demand location is not necessarily directly defined; instead, there are several paths from the supply location to the demand location, among these some yield maximal surplus. Therefore, both the optimal assignment problem and the shortest path problem are instances of the optimal flow problem; these instances are representative in the sense that any optimal flow problem may be decomposed into an assignment problem and a number of shortest path problems. We will show how to easily compute these problems using linear programming.

Chapter 9 offers a selection of applications to economics: partial identification in econometrics in section 9.1, inversion of demand systems in section 9.2, computation of hedonic equilibria in section 9.3, identification via vector quantile methods in section 9.4, quantile regression in section 9.5, multidimensional screening in section 9.6, and pricing of financial derivatives in section 9.7.

Chapter 10 concludes with perspectives on computation, duality, and equilibrium.

1.7 NOTATION AND CONVENTIONS

Throughout this text we have tried to strike a good balance between mathematical precision and ease of exposition. A *probability measure* will always mean a Borel probability measure; a *set* will always mean a measurable set; a *continuous probability* or *continuous distribution* means a probability measure which is absolutely continuous with respect to the Lebesgue measure; a *convex* function means a convex function which is not identically $+\infty$.

Usual abbreviations will be used: *c.d.f.* means the cumulative distribution function; *p.d.f.* means probability density function; *a.s.* means almost surely; depending on the context, *s.t.* means such that or subject to.

Given a smooth function $f : \mathbb{R}^d \to \mathbb{R}$, the gradient of f at x, denoted $\nabla f(x)$, is the vector of partial derivatives $(\partial f(x)/\partial x_1, \dots, \partial f(x)/\partial x_d)$. Given a function $f : \mathbb{R}^k \to \mathbb{R}^l$, $Df(x)$ is the Jacobian matrix of f at x, that is, the matrix of partial derivatives $(\partial f^i(x)/\partial x^j)$, $1 \le i \le l$, $1 \le j \le k$. The Dirac mass at x_0, denoted δ_{x_0}, is the probability distribution whose c.d.f. is $F(x) = 1\{x_0 \le x\}$. The notation $A \succeq_{spd} B$ means $A - B$ is symmetric, positive semidefinite. Given a compact set C of $\mathbb{R}^d$, the notation $\mathcal{U}(C)$ denotes the uniform distribution on C, which is simply denoted $\mathcal{U}$ when $C = [0, 1]$ is the unit interval. The set $L^1(P)$ is the set of functions which are integrable with respect to P. $X \sim P$ means X has distribution P. Perhaps less standard notation is $\mathcal{M}(P, Q)$, which denotes the set of couplings of P and Q, which is the set of probability measures π such that if $(X, Y) \sim \pi$, then $X \sim P$ and $Y \sim Q$.

Throughout this book, we shall try to have consistent notation, which we now summarize. Some exceptions are made in chapter 9, where we sometimes choose to use notation that is more traditional in the given application.

Agents: The types or attributes of workers are denoted by $x \in \mathcal{X}$, and the types or attributes of firms (or jobs) are denoted by $y \in \mathcal{Y}$. Whenever needed, i indexes individual workers, and j indexes individual firms.

Probabilities: A probability measure on $\mathcal{X}$ is denoted by P and on $\mathcal{Y}$ by Q. The probability measure of observing pair (x, y) is denoted by $\pi \in \mathcal{M}(P, Q)$; according to the context (continuous or discrete), this probability measure may be identified with its p.d.f. or with its probability mass function in the discrete case.

Utilities: The pretransfer surplus of a type x when matched with a type y is denoted by $\alpha(x, y)$, and the pretransfer surplus of a type y when matched with a type x is denoted by $\gamma(x, y)$. The matching surplus of a couple (x, y) is denoted by $\Phi(x, y) = \alpha(x, y) + \gamma(x, y)$. The equilibrium monetary transfer (wage) from y to x within pair (x, y) is denoted by $w(x, y)$. The posttransfer indirect surplus of a type x is denoted by $u(x)$, and the posttransfer indirect surplus of a type y is denoted by $v(y)$.

Whenever types x and y are discrete, we will prefer to use subscript notation, that is, Φ_{xy} instead of $\Phi(x, y)$.

═══2═══

Monge–Kantorovich Theory

Consider the problem of assigning a possibly infinite number of workers and firms. Each worker should work for one firm, and each firm should hire one worker. Workers and firms have heterogenous characteristics; let $x \in \mathcal{X}$ and $y \in \mathcal{Y}$ be the vectors of characteristics of workers and firms respectively. In all generality, $\mathcal{X}$ and $\mathcal{Y}$ are *Polish spaces*,[1] but, for all practical purposes in these notes, they will be either open or closed subsets of $\mathbb{R}^d$. It is assumed that workers and firms are in equal mass, so that the total mass of workers is normalized to 1, as is the total mass of firms. The distribution of workers' types is P, and the distribution of the firms' types is Q, where P and Q are probability measures on $\mathcal{X}$ and $\mathcal{Y}$. Let us first describe what an assignment might look like; for this, we need to introduce the notion of couplings.

2.1 COUPLINGS

A *coupling* determines which workers are assigned to which firms. If we had a finite number of workers and firms, we would need to count the number of workers of a given type matched with firms of a given type. More generally, a coupling will be defined as the probability measure π of the occurrence of worker–firm pairs. If $(X, Y) \sim \pi$ is the joint random pair, then $X \sim P$ and $Y \sim Q$, where $X \sim P$ means X *has distribution* P. In other words, the first *margin* of π should be P, while its second margin should be Q. This motivates the following definition.

DEFINITION 2.1. *The set of couplings of probability distributions P and Q is the set of probability distributions over $\mathcal{X} \times \mathcal{Y}$ with first and second margins P and Q. This set is denoted $\mathcal{M}(P, Q)$. That is, a probability measure π over $\mathcal{X} \times \mathcal{Y}$ is in $\mathcal{M}(P, Q)$ if and only if*

$$\pi(A \times \mathcal{Y}) = P(A) \quad and \quad \pi(\mathcal{X} \times B) = Q(B)$$

---

[1] A Polish space is a separable, completely metrizable, topological space.

holds for every subset A of $\mathcal{X}$ and B of $\mathcal{Y}$. By extension, a random pair $(X, Y) \sim \pi$, where $\pi \in \mathcal{M}(P, Q)$, will also be called a coupling of P and Q.

Let us discuss some particular couplings. The simplest of them is arguably the *independent coupling*, sometimes rather improperly called *random matching*: it consists of taking $\pi (A \times B) = P(A) Q(B)$, so that, if $(X, Y) \sim \pi$, then $X \sim P$ and $Y \sim Q$ are independent. The existence of this coupling establishes that the set of couplings $\mathcal{M}(P, Q)$ is nonempty. However, it is not very relevant from an economic point of view; in our worker–firm assignment example, we would not expect to see workers randomly matched to firms.

On the contrary, it will be of interest to consider couplings (X, Y) such that Y is a deterministic function of X; that is, $Y = T(X)$. In our worker–firm example, this assumes that every worker of type x will get assigned to the same type of firm, $T(x) \in \mathcal{Y}$. This type of coupling is called a *pure assignment*, or *Monge coupling*. The constraint on T that ensures that the distribution of $(X, T(X))$ is a coupling of P and Q is equivalent to any of the following five conditions:

(i) If $X \sim P$, then $T(X) \sim Q$.
(ii) Equality $P\left(T^{-1}(B)\right) = Q(B)$ holds for every subset B of $\mathcal{Y}$.
(iii) The coupling defined by $\pi(x, y) := P(x)\delta(y - T(x))$ is in $\mathcal{M}(P, Q)$.
(iv) For any $\varphi \in L^1(Q)$,

$$\mathbb{E}_P\left[\varphi(T(X))\right] = \mathbb{E}_Q\left[\varphi(Y)\right]. \tag{2.1}$$

(v) When P and Q have respective densities f_P and f_Q, and when T is smooth, the *Monge–Ampère equation* holds:

$$f_P(x) = |\det DT(x)| f_Q(T(x)), \tag{2.2}$$

which is nothing other than a multivariate change of variables in (2.1).

These equivalent conditions are denoted by

$$T\#P = Q, \tag{2.3}$$

where $T\#P$ is the distribution of $T(X)$ when $X \sim P$, called the *push-forward* of probability distribution P by map T. In the probability literature, expression (2.3) is sometimes denoted $PT^{-1} = Q$.

There are, of course, many couplings that are not Monge couplings. In general, one can equivalently characterize a coupling by a family of conditional probability distributions, or *Markov kernels*, $\pi(dy|x)$ such that

$$\int_{\mathcal{X}} \pi(B|x) \, dP(x) = Q(B)$$

for every subset B of $\mathcal{Y}$.

Figure 2.1: Gaspard Monge (1746–1818). © Collections École Polytechnique (Palaiseau, France).

2.2 OPTIMAL COUPLINGS

Assume now that if worker x works for firm y, this generates a quantity of output $\Phi(x, y)$, measured in some monetary unit. Consider the problem of a social planner, who decides which workers to assign to which firms, and wants to maximize the total output. This is precisely the question addressed by the Monge–Kantorovich problem.

The *Monge problem* consists of looking among all the *pure* assignments of firms to workers for the one that maximizes $\mathbb{E}_P\left[\Phi\left(X, T(X)\right)\right]$, the average overall surplus. That is,

$$\max_{T(\cdot)} \mathbb{E}_P\left[\Phi\left(X, T(X)\right)\right]$$
$$\text{s.t. } T\#P = Q. \tag{2.4}$$

Monge described the problem under this form in the early nineteenth century, taking a distance cost, that is, $\Phi(x, y) = -|x - y|$. Despite making interesting geometric observations on optimality conditions, Monge did not succeed in providing a full solution. After him, the problem remained a well-known challenge for nineteenth-century mathematicians. To make progress, a big idea was needed, and it took a century and a half before *linear programming relaxation* arrived with the work of Kantorovich in the 1940s. Rather than imposing every worker of type x to work with firm $T(x)$, Kantorovich introduced the possibility of randomization: instead of maximizing over all deterministic assignment maps $T(x)$ as in (2.4), one can maximize over the conditional probabilities (Markov kernels) of assigning worker x to firm y. Thus, $\Phi(x, T(x))$ in (2.4) is replaced by $\mathbb{E}_\pi[\Phi(x, Y)|X = x]$, where $\pi \in \mathcal{M}(P, Q)$ is a coupling of probabilities P and Q which is not necessarily pure. This led Kantorovich to formulate what is now called the *Kantorovich problem*,

$$\max_{\pi \in \mathcal{M}(P,Q)} \mathbb{E}_\pi[\Phi(X, Y)], \tag{2.5}$$

which, compared to the Monge problem (2.4), has several important advantages. First, it is a linear programming problem, as the objective function is linear with respect to the optimization variable π, and so are the constraints. The linear programming nature will lead to a powerful duality theory described in the next paragraph. Second, the Kantorovich problem always has a solution π, which is not always the case with the Monge problem, which may have no solution. Third, in a number of important cases which we will study in detail, it will however be the case that the solutions to the Monge and Kantorovich problems coincide, namely, that the solution to the Kantorovich problem is a pure assignment. In this case the Kantorovich relaxation offers a solution to the Monge problem.

2.3 MONGE–KANTOROVICH DUALITY

Without further delay, we now state the Monge–Kantorovich theorem, also known as the Monge–Kantorovich duality theorem. This fundamental result will be the cornerstone of much of this book.

THEOREM 2.2 (Monge–Kantorovich duality). *Let $\mathcal{X}$ and $\mathcal{Y}$ be two Banach spaces, and let P and Q be two probability measures on $\mathcal{X}$ and $\mathcal{Y}$ respectively. Let $\Phi : \mathcal{X} \times \mathcal{Y} \to \mathbb{R} \cup \{-\infty\}$ be an upper semicontinuous surplus function bounded from above. Then,*

(i) the value of the primal Monge–Kantorovich problem

$$\sup_{\pi \in \mathcal{M}(P,Q)} \mathbb{E}_\pi[\Phi(X, Y)] \tag{2.6}$$

Figure 2.2: Leonid Kantorovich (1912–1986).

coincides with the value of the dual

$$\inf_{u,v} \mathbb{E}_P\left[u(X)\right] + \mathbb{E}_Q\left[v(Y)\right]$$

$$\text{s.t. } u(x) + v(y) \geq \Phi(x,y),$$

(2.7)

where the infimum is over measurable and integrable functions u and v, and the inequality constraint should be satisfied for almost every x and almost every y (all these statements are respective to measures P and Q);
(ii) an optimal solution π to problem (2.6) exists.

This theorem has not been stated here in its most general possible form. As mentioned above, the result holds if $\mathcal{X}$ and $\mathcal{Y}$ are Polish spaces instead of being Banach spaces. However, for practical purposes, we will never need this level of generality, which is the reason why we did not impose it in the statement of theorem 2.2. Also, instead of assuming Φ is bounded from above, it is sufficient to assume $\Phi(x,y) \leq a(x) + b(y)$ for $a \in L^1(P)$ and $b \in L^1(Q)$, two upper semicontinuous functions. See [149, theorem 5.10] and theorem 7.6.

Note that the expressions of the values of problems (2.6) and (2.7) provide two ways of computing the total surplus. The expression of the value of problem (2.6) provides a breakdown of the surplus by pairs. The total surplus is computed as the sum over the possible pairs of the pairwise surplus integrated against the distribution of mass of these pairs. In contrast, the value of problem (2.7) offers a breakdown of the total surplus at the individual level. Indeed, in the next paragraph, $u(x)$ and $v(y)$ will find an interpretation as the equilibrium payoffs that worker x and firm y get at equilibrium; under this interpretation, the total surplus is written in problem (2.7) as the sum of the worker's equilibrium surplus integrated against the distribution in the population, plus the sum of the firm's surpluses integrated against their distribution.

As noted above, theorem 2.2 is a result in infinite-dimensional linear programming. The primal optimization variable π is a probability measure, hence (as soon as the cardinality of $\mathcal{X}$ or $\mathcal{Y}$ is infinite) an infinite-dimensional vector. Similarly, the dual optimization variables are integrable functions u and v, which are also infinite-dimensional vectors as soon as the cardinality of $\mathcal{X}$ or $\mathcal{Y}$ respectively is infinite. In general, there are fewer results for infinite-dimensional linear programming than for finite-dimensional linear programming. *Weak duality*, which states the value of the dual (i.e., the minimization problem) is always weakly greater than the value of the primal problem (i.e., the maximization problem), still holds. However, *strong duality*, which states that the values of these two problems coincide, does not hold true in general. Also, in contrast to the finite-dimensional case, the primal or the dual problem may have a finite value without having an optimal solution. However, theorem 2.2 indicates that the Monge–Kantorovich problem is a somewhat "well-behaved" infinite-dimensional linear programming problem. Part (i) of this result implies that a strong duality holds: the values of the primal and the dual problems coincide. Part (ii) ensures that the primal problem has an optimal solution. Note, however, that the theorem does not make a statement about the existence of solutions (u, v) to the dual problem (2.7), which, when they exist, are called *Kantorovich potentials*. In fact, under the assumptions of theorem 2.2, a solution to the dual problem does not necessarily exist. We shall see in chapter 7 that more restrictive assumptions on the surplus function will yield the existence of solutions to the dual problem, and that in some cases, properties of the solutions to the dual problem will lead to conclusions on the existence of a solution to the Monge problem.

It is easy to show a part of theorem 2.2(i), namely, the weak duality inequality. Consider π satisfying the constraints of problem (2.6), and a pair of integrable functions u and v satisfying the constraints of problem (2.7). Then $\pi \in \mathcal{M}(P, Q)$ and $u(x) + v(y) \geq \Phi(x, y)$. As a result, for $(X, Y) \sim \pi$, one has $u(X) + v(Y) \geq \Phi(X, Y)$ π-almost surely. Thus, taking expectations with respect to π, and using the fact that $\mathbb{E}_\pi [u(X)] = \mathbb{E}_P [u(X)]$ and $\mathbb{E}_\pi [v(Y)] = \mathbb{E}_Q [v(Y)]$, it follows that

$$\mathbb{E}_P [u(X)] + \mathbb{E}_Q [v(Y)] \geq \mathbb{E}_\pi [\Phi(X, Y)].$$

Taking the infimum of the left-hand side and the supremum of the right-hand side yields the easy part of the Monge–Kantorovich theorem,

$$\inf_{u(x)+v(y)\geq\Phi(x,y)} \mathbb{E}_P\left[u(X)\right] + \mathbb{E}_Q\left[v(Y)\right] \geq \sup_{\pi\in\mathcal{M}(P,Q)} \mathbb{E}_\pi\left[\Phi(X,Y)\right].$$

However, the fact that this inequality is actually an equality, namely, that *strong* duality holds, is the difficult part of theorem 2.2(i). It is based on the abstract Fenchel–Rockafellar duality, which is itself a consequence of the Hahn–Banach theorem of separation of convex sets. Part (ii) of the result (the existence of an optimal π) is not a very difficult result, but requires a bit of topology. It comes from the compactness of $\mathcal{M}(P,Q)$, which is a consequence of Prokhorov's theorem. We shall content ourselves with giving a sketch of the proof of theorem 2.2(i), which is given in full in [148, section 1.1], to which we refer the reader for the complete proof.

SKETCH OF PROOF OF THEOREM 2.2. The value of the primal problem 2.6 can be rewritten as

$$\sup_{\pi\in\mathcal{M}^+} \int \Phi(x,y)\, d\pi(x,y) + A_{P,Q}(\pi),$$

where $\mathcal{M}^+$ is the set of positive measures over $\mathcal{X}\times\mathcal{Y}$ (not necessarily of total mass 1, and not necessarily with fixed marginals), and $A_{P,Q}$ should be such that

$$A_{P,Q}(\pi) = \begin{cases} 0 & \text{if } \pi\in\mathcal{M}(P,Q), \\ -\infty & \text{otherwise.} \end{cases}$$

One can take

$$A_{P,Q}(\pi) = \inf_{u,v} \int u(x)\, dP(x) + \int v(y)\, dQ(y) - \int (u(x)+v(y))\, d\pi(x,y),$$

so that the value of the primal problem becomes

$$\sup_{\pi\in\mathcal{M}^+} \inf_{u,v} \left\{ \int (\Phi(x,y) - (u(x)+v(y)))\, d\pi(x,y) \right.$$
$$\left. + \int u(x)\, dP(x) + \int v(y)\, dQ(y) \right\}.$$

It is the case here that sup inf = inf sup (this fact will be admitted without a proof), which yields

$$\inf_{u,v} \int u(x)\, dP(x) + \int v(y)\, dQ(y) + B_\Phi(u,v),$$

where $B_\Phi(u, v) = \sup_{\pi \in \mathcal{M}^+} \int (\Phi(x, y) - (u(x) + v(y)))\, d\pi(x, y)$, so that

$$B_\Phi(u, v) = \begin{cases} 0 & \text{if } u(x) + v(y) \geq \Phi(x, y) \text{ for all } x \text{ and } y, \\ +\infty & \text{otherwise;} \end{cases}$$

thus the value of the problem can be rewritten as (2.7). As noted above, this argument is only a rough sketch; the minimax principle which we invoked when inverting the sup and the inf needs to be carefully established, and the spaces in which the functions u and v and the measures π live need to be made precise. Again, the reader will find the rigorous argument in Villani's text. $\quad\square$

2.4 EQUILIBRIUM

Let us now discuss the economic interpretation of this result, and especially the dual problem. Remember, the assumptions of theorem 2.2 do not imply that a pair (u, v) of solutions to the dual problem (2.7) exists. However, consider a pair (u, v) of solutions to the dual problem (2.7), assuming it exists. We shall argue that $u(x)$ can be interpreted as the equilibrium wage of worker x, while $v(y)$ can be interpreted as the equilibrium profit of firm y.

PROPOSITION 2.3. *If (u, v) is a solution to the dual of the Kantorovich problem, then we can always redefine u and v so that they take the value $+\infty$ outside the supports of P and Q, respectively. In this case,*

$$u(x) = \sup_y \left(\Phi(x, y) - v(y) \right), \tag{2.8}$$

$$v(y) = \sup_x \left(\Phi(x, y) - u(x) \right) \tag{2.9}$$

should hold almost surely with respect to the probabilities P and Q, respectively.

PROOF. The constraint in (2.7) implies $u(x) + v(y) \geq \Phi(x, y)$, thus

$$v(y) \geq \sup_x \left(\Phi(x, y) - u(x) \right),$$

but if the latter inequality was to hold strictly on a set with positive Q measure, one could strictly improve on the dual objective function $\mathbb{E}_P[u(X)] + \mathbb{E}_Q[v(Y)]$, and contradict optimality of (u, v). Hence the inequality is actually an equality, and (2.8) holds. $\quad\square$

Expression (2.9) stands therefore for the problem of a firm choosing optimally the worker it will hire. It implies that $u(x)$ can be interpreted as the market wage of worker x, and $v(y)$ as the indirect surplus of firm y: the firm will not hire worker x unless the profit (net of wage) of hiring x coincides with $v(y)$, which is the profit that the firm knows it can obtain. Expression (2.8) expresses symmetrically the worker's problem of choosing a firm optimally. As a result, theorem 2.2 can be interpreted as a welfare theorem: the solution of the central planner (2.6), coincides with the solution of the decentralized equilibrium, given by (2.7). Moreover, the decentralized solution is not only efficient in the sense of Pareto, it is also efficient in a utilitarian sense, in that it maximizes the sum of the surplus $\Phi(x, y)$ over the mass of pairs (x, y), among the feasible distribution of pairs.

Of course, an important indeterminacy remains in the dual problem: again, if (u, v) is a solution of the dual, and if $c \in \mathbb{R}$, then $(u - c, v + c)$ is also a solution. In other words, if $u(x)$ is an equilibrium wage curve, then so is $u(x) + c$. This is due to the fact that we have not specified the outside option. In this model, all agents are forced to match, so there is no external benchmark to fix the value of c. In order to fully close the model, we should specify this outside option. For instance, we may assume that workers get a payoff equal to 0 if they don't work, and that there are infinitesimally more workers than firms. In this case, the reservation wage will force $u(x) \geq 0$, as working for a negative wage would be irrational. Furthermore, competition between workers (who slightly outnumber firms) will lead to the fact that the least paid worker is actually indifferent between working and not working, and hence receives a wage equal to 0. This will fully close the model and will determine the constant c.

2.5 A PREVIEW OF APPLICATIONS

As stated in the introduction, the Monge–Kantorovich problem is the relevant structure for a number of diverse topics in economics. Let us begin with some examples that are direct instances of the Monge–Kantorovich problem. Our first example deals with the worker–firm assignment described in the present chapter; the only difference is that we consider the particular case where there are finite numbers of workers and firms.

Example 2.1 (Optimal discrete assignments). *Consider the worker–firm assignment problem of chapter 3. Then $\mathcal{X}$ and $\mathcal{Y}$ are discrete and finite, and the duality in theorem 3.1 can be seen as a particular case of theorem 2.2. In this case, the support of P is finite, and denoted $\{x_1, \ldots, x_N\}$; similarly, the support of Q is denoted $\{y_1, \ldots, y_M\}$. Let p_i be the mass given by P to x_i and q_j be the mass given by Q to y_j, so that*

$$P = \sum_{i=1}^{N} p_i \delta_{x_i} \quad and \quad Q = \sum_{j=1}^{M} q_j \delta_{y_j},$$

where δ_x denotes the Dirac measure at x, defined in section 1.7, and $\Phi_{ij} =$ $\Phi\left(x_i, y_j\right)$. In this case, it is not hard to see that the solution to (2.6) is of the form $\pi = \sum_{ij} \pi_{ij} \delta_{x_i} \delta_{y_j}$, and the solution to (2.7) is such that $u_i = u\left(x_i\right)$ and $v_j = v\left(y_j\right)$, where $\left(\pi_{ij}\right)$ and $\left(u_i, v_j\right)$ are solutions to (3.3) and (3.4), respectively.

Our second example is a celebrated application of assignment models to the analysis of the behavior of populations of men and women who seek to match on the marriage market.

Example 2.2 (The Becker model of matching). *Becker [11] considers a model of the marriage market where $x \in \mathcal{X}$ and $y \in \mathcal{Y}$ are scalar "ability indices" of men and women, respectively, and the joint matching surplus to be shared among a man x and a woman y is $\Phi(x, y)$. It is usually assumed that $\partial^2 \Phi(x, y) / \partial x \, \partial y \geq 0$. This property is called positive assortative matching and implies that high types match with high types. There are many connections with quantile and probability transforms, copula theory, and estimation under shape restriction, investigated in chapter 4.*

Our third example models the way consumers choose certain differentiated goods, such as cars, in the space of characteristics.

Example 2.3 (Characteristics-based demand). *We consider Hotelling's and Lancaster's characteristics-based approaches to consumer demand (also known as address models). Assume that a consumer's vector of characteristics is x, which is distributed along P on $\mathcal{X}$, which is a subset of $\mathbb{R}^d$. There are M products to choose from, indexed by j. Product j's characteristics are a vector $y_j \in \mathcal{Y}$, where $\mathcal{Y}$ is a finite subset of $\mathbb{R}^d$. In addition, the price of product j is π_j. The utility of consumer x conditional on choosing product j is*

$$U(x) = - \left\| x - y_j \right\|^2 - \pi_j;$$

hence it decreases when the price of the good increases and when the characteristics of the good get further away from the consumer's. This model, along with its connections with Voronoi cells and power diagrams, will be investigated in chapter 5.

Our next examples require more abstraction. They consist of showing that some classical problems in economics and econometrics can be reformulated as problems of the Monge–Kantorovich type. The first example of this kind consists of arguing that the problem of nonparametrically inverting a discrete choice model (also known as the problem of conditional choice probability inversion) is actually a Monge–Kantorovich problem.

Example 2.4 (Inversion of discrete choice models). *We consider a discrete choice model. Consumer heterogeneity is represented by $\varepsilon \sim P$, which takes values over $\mathcal{X}$, a subset of $\mathbb{R}^d$, with distribution P. There are d alternatives* *to choose from, labeled $y \in \mathcal{Y} = \{1, \ldots, d\}$. The surplus of consumer ε choosing alternative y is $w_y + \varepsilon_y$, hence the consumer's problem is*

$$\max_{y} \{w_y + \varepsilon_y\}.$$

We are interested in the inversion of this model: Given the observation of the probability q_y of choosing each y, how can we identify w (of course, up to an additive constant)? In section 9.2, we shall see that this problem is solved by the fact that $v_y = -w_y$ is the solution to problem (2.7), with $\Phi(\varepsilon, y) = \varepsilon_y$.

The following example deals with a cousin of the matching model, called the hedonic model. In hedonic models, people do not really care about the identity of their partners, but they care about what they achieve jointly with their partners. A consumer may not care about the identity of the hairdresser, but instead cares about the quality of the haircut.

Example 2.5 (Hedonic models). *We consider the problem of "hedonic equilibrium." Assume that a population of producers is distributed according to a distribution P of characteristics x over a subset $\mathcal{X}$ of $\mathbb{R}^d$. The producers have the option to produce a unit of good in various qualities z in a subset $\mathcal{Z}$ of $\mathbb{R}^d$. The quality z has an endogenous Walrasian price $w(z)$. It is assumed that if producer x produces quality z, the surplus will be $\alpha(x, z) + w(z)$. Similarly, the distribution of consumer characteristics y over $\mathbb{R}^d$ is $\mathcal{Y}$, and a consumer y consuming quality z gets surplus $\gamma(y, z) - w(z)$. In section 9.3, we shall see that the equilibrium probability $\pi(x, y)$ that a consumer y buys from a producer x is the solution to problem (2.6), where*

$$\Phi(x, y) = \max_{z} \{\alpha(x, z) + \gamma(y, z)\}$$

and the set of equilibrium prices $w(\cdot)$ can be easily determined from the solutions to (2.7).

Finally, we argue that some identification problems in principal–agent models can be reformulated as Monge–Kantorovich problems.

Example 2.6 (Principal–agent problems). *We consider a basic principal–agent model with possibly multivariate characteristics. Consider an agent of type $x \in \mathcal{X}$, where $\mathcal{X}$ is a subset of $\mathbb{R}^d$. Assume that the types of agents follow a probability distribution P over $\mathcal{X}$. Based on the type x' announced by the agent, the principal decides on an outcome $y = T(x') \in \mathcal{Y}$, where $\mathcal{Y}$ is also a subset of $\mathbb{R}^d$, and on a payment $v(y)$ made by the agent, so that agent x's utility of announcing type x' is $\Phi(x, T(x')) - v(T(x'))$. Then T is implementable in dominant strategy if and only if there exists a payment schedule $v(\cdot)$ such that*

$$T(x) \in \arg\max_{y} \{\Phi(x, y) - v(y)\}.$$

As we shall see in section 9.6, this model can be reformulated as an optimal transportation problem. The key result will be that T is implementable if and only if for $X \sim P$, the distribution π of $(X, T(X))$ is the solution to problem (2.6). In particular, if the observations consist of (X, Y), where X is the type announced by agent X and Y is the outcome, then v can be identified by means of problem (2.7).

2.6 EXERCISES

Exercise 2.1 (M). ***Fixed effects***. Show that the set of solutions to the primal Monge–Kantorovich problem (2.6) remains the same if $\Phi(x, y)$ is replaced by $\Phi(x, y) + a(x) + b(y)$. What is the effect on the primal and dual solutions?

Exercise 2.2 (M). ***The Wasserstein distance***. Assume $\mathcal{X} = \mathcal{Y} = \mathbb{R}^d$, and that P and Q have second moments. Show that the value of

$$B = \max_{\pi \in \mathcal{M}(P,Q)} \mathbb{E}_\pi \left[X'Y \right] \tag{2.10}$$

can be related to the value of

$$W = \min_{\pi \in \mathcal{M}(P,Q)} \mathbb{E}_\pi \left[|X - Y|^2 \right], \tag{2.11}$$

up to constants to be characterized, and show that the optimal π is the same in the two problems. The value of program (2.11) is called the squared Wasserstein distance between P and Q. Write down the dual problem associated with both problems (2.10) and (2.11) and relate the solution to these two problems.

Exercise 2.3 (M). ***On the unimportance of the constant***. Show that the dual Monge–Kantorovich problem (2.7) also can be rewritten as

$$\inf_{u, v} \mathbb{E}_Q \left[v(Y) \right]$$

$$\text{s.t. } u(x) + v(y) \geq \Phi(x, y), \tag{2.12}$$

$$\mathbb{E}_P \left[u(X) \right] = 0.$$

Using an informal minimax formulation, compute the dual of the latter program and interpret the value of the shadow price of its bottom constraint.

Exercise 2.4 (E). ***No interaction***. Assume $\Phi(x, y) = a(x) + b(y)$. Characterize the solutions to the primal and dual Monge–Kantorovich problems and provide an economic interpretation.

Exercise 2.5 (E). ***Walrasian wages***. Assume that worker characteristics are drawn from a population distribution P, and firm characteristics are drawn from *a population distribution Q. Assume that worker x has utility surplus $\alpha(x, y) + w$ of working for firm y at wage w, and firm y has surplus $\gamma(x, y) - w$. Let π be the equilibrium assignment, and w(x, y) be the equilibrium wage.*

(i) *Show that the optimum assignment of firms to workers is a solution to problem (2.6), with $\Phi(x, y) = \alpha(x, y) + \gamma(x, y)$.*

(ii) *Assuming P, Q, and Φ are such that both problems (2.6) and (2.7) have solutions, show that w(x, y) is an equilibrium wage if and only if there is a solution (u, v) of problem (2.7) such that for every x and y in the support of P and Q,*

$$\gamma(x, y) - v(y) \leq w(x, y) \leq u(x) - \alpha(x, y).$$

Exercise 2.6 (E). **The Becker–Coase theorem.** *Assume that a new bill requires landlords to pay the broker's fees, which were previously customarily paid by tenants. Rent prices are not controlled (and thus are adjusted by the market), and it is assumed that there is a slight excess supply of rental houses in the market under consideration. Using the formalism developed in this chapter, and more particularly the result of exercise 2.5, argue why the bill may be inefficient.*

2.7 REFERENCES AND NOTES

Part of our exposition in chapter 2 is inspired by Villani [148]; see also Villani [149]. We refer to Vershik [147] for an informative historical account. Monge's 1781 treatise [109], *Mémoire sur la théorie des déblais et des remblais*, appeared in French, while Kantorovich's results appeared in Russian in [82, 83]. Kantorovich later realized the connection with Monge's problem in [84]. A few years afterward, Koopmans published a series of papers on the "transportation problem" in connection with linear programming; see [89, 90]. Both Kantorovich and Koopmans are regarded as pioneers in the development of linear programming and its economic applications; they shared the Nobel Memorial Prize in 1975 for "their contribution to the theory of optimal allocation of resources." Shapley and Shubik [138], Becker [11], and Gretsky, Ostroy, and Zame [73] proposed variants of theorem 2.2 which allow for unassigned agents. A reference for the Becker–Coase theorem which is the focus of exercise 2.6 is Becker [12, p. 424].

—3—

The Discrete Optimal Assignment Problem

We still consider the problem of assigning workers to jobs as in the previous chapter, but we now assume that the type spaces $\mathcal{X}$ and $\mathcal{Y}$ are finite. Thus we set $\mathcal{X} = \{1, \ldots, N\}$, and $\mathcal{Y} = \{1, \ldots, M\}$, so that workers have types $x \in \{1, \ldots, N\}$ and jobs have types $y \in \{1, \ldots, M\}$. The total mass of workers and jobs is normalized to 1. The mass of workers of type x is p_x; the mass of jobs of type y is q_y. Let π_{xy} be the mass of workers of type x assigned to jobs of type y. The conditions on the assignment matrix π are that every worker is occupied and every job is filled, that is,

$$\sum_{y=1}^{M} \pi_{xy} = p_x \quad \text{and} \quad \sum_{x=1}^{N} \pi_{xy} = q_y. \tag{3.1}$$

Note that this formulation implicitly allows workers to multitask, that is, it allows for $\pi_{xy} > 0$ and $\pi_{xy'} > 0$ to hold simultaneously with $y \neq y'$. In section 3.3, we shall examine what happens when we forbid this possibility and impose that workers of a given type work for only one type of job.

The economic value created when assigning worker x to job y is Φ_{xy}. As before, it is assumed that the total value originated is the sum of the value created by each pair, that is, $\sum_{xy} \pi_{xy}\Phi_{xy}$. The value of the optimal assignment is the maximum of the latter subject to constraints (3.1). The problem is therefore

$$\max_{\pi \geq 0} \sum_{xy} \pi_{xy}\Phi_{xy} \tag{3.2}$$

$$\text{s.t. (3.1)},$$

and it is now a finite-dimensional linear programming problem. In the rest of this chapter, we shall first see how the general results viewed in chapter 2 particularize to this case. Then we shall move on to practical ways of numerically solving problem (3.2) on a computer.

3.1 DUALITY

In this section, we examine what becomes of theorem 2.2 in the current setting when $\mathcal{X}$ and $\mathcal{Y}$ are finite. Although straightforward, the following result is of considerable importance as most numerical methods we will use for the computation of the Monge–Kantorovich problem will involve finite-dimensional approximations.

THEOREM 3.1.   (i) *The value of the primal problem*

$$\max_{\pi_{xy} \geq 0} \sum_{xy} \pi_{xy} \Phi_{xy}$$

$$\text{s.t.} \sum_{y=1}^{M} \pi_{xy} = p_x \quad and \quad \sum_{x=1}^{N} \pi_{xy} = q_y \tag{3.3}$$

coincides with the value of the dual problem

$$\min_{u,v} \sum_{x=1}^{N} p_x u_x + \sum_{y=1}^{M} q_y v_y \tag{3.4}$$

$$\text{s.t.} \ u_x + v_y \geq \Phi_{xy}.$$

(ii) *Both the primal and the dual problems have optimal solutions. If π is a solution to the primal problem and (u, v) a solution to the dual problem, then by complementary slackness,*

$$\pi_{xy} > 0 \quad implies \quad u_x + v_y = \Phi_{xy}. \tag{3.5}$$

(iii) *If (u, v) is a solution to the dual problem, then*

$$u_x = \max_{y \in \{1,\dots,M\}} \left\{ \Phi_{xy} - v_y \right\} \quad and \quad v_y = \max_{x \in \{1,\dots,N\}} \left\{ \Phi_{xy} - u_x \right\}. \tag{3.6}$$

Before we give the proof of this result, let us briefly discuss it. The only novel item which is evidenced by the finite-dimensional formulation is point (ii): complementary slackness. This remark implies that whenever there is a match between x and y, then $u_x + v_y$ should be an actual breakdown of the surplus Φ_{xy}. We will interpret this by the means of feasibility, in the next section. Indeed, u_x and v_y will be interpreted as payoffs that x and y get respectively in a stable outcome.

The optimization problems involved are now finite-dimensional, so the result follows directly from application of standard results on linear programming duality. However, it may be interesting for some readers to recall how the latter are themselves a consequence of the minimax theorem.

PROOF OF THEOREM 3.1.   (i) The value of the primal problem (3.3) can be written as $\max_{\pi \geq 0} \min_{u,v} S(\pi, u, v)$, where

$$S(\pi, u, v) := \sum_{xy} \pi_{xy} \Phi_{xy} + \sum_{x=1}^{N} u_x \left(p_x - \sum_{y=1}^{M} \pi_{xy} \right) + \sum_{y=1}^{M} v_y \left(q_y - \sum_{x=1}^{N} \pi_{xy} \right),$$

but by the minimax theorem B.1, this value is equal to $\min_{u,v} \max_{\pi \geq 0} S(\pi, u, v)$, which is the value of the dual problem (3.4).

(ii) This follows by noting that, for a primal solution π and a dual solution (u, v), then $S(\pi, u, v) = \sum_{xy} \pi_{xy} \Phi_{xy}$.

(iii) This result has already been argued in proposition 2.3.   □

3.2 STABILITY

There is an interesting interpretation of the previous duality results in terms of *stability*. Consider a game where workers and job-offering firms have to form pairs. If worker x pairs up with job y, they together generate output Φ_{xy}, which they have to share between them. An outcome (π_{xy}, u_x, v_y) is the specification of the mass π_{xy} of formed pairs of type xy, along with the payoff u_x that a worker of type x gets, as well as the payoff v_y that firm y gets. It is assumed that if a worker or a firm are unassigned, then they will get utility $-\infty$; in other words, individuals in this model will match at any price.[1]

Stability is the combination of two properties: feasibility and the absence of a blocking pair, which we now discuss. A basic requirement for an outcome is to be *feasible*, namely, that the total quantity of output generated is equal to the total quantity of output redistributed to workers and firms. This is expressed as

$$\sum_{xy} \pi_{xy} \Phi_{xy} = \sum_{x=1}^{N} p_x u_x + \sum_{y=1}^{M} q_y v_y, \tag{3.7}$$

which shows how the total output generated at the pairwise level (left-hand side expression) is redistributed at the individual level (right-hand side expression). The next important concept is the *absence of a blocking pair*. Consider an outcome (π, u, v), and assume that there is some worker of type x and some firm of type y such that $u_x + v_y < \Phi_{xy}$. Then, this worker and this firm will have an incentive to resign from their current arrangement, and

---

[1] In contrast, the standard assignment model of Shapley and Shubik [138] assumes that individuals have a finite reservation utility, which implies $u, v \geq 0$. This is the only difference between their model and the one presented here, albeit a significant one.

match together; they may do so in a way which guarantees strictly more than what they have under the current outcome, as x may obtain $u_x + \varepsilon$, while y may obtain $v_y + \varepsilon$, with $\varepsilon = \frac{1}{2}(\Phi_{xy} - u_x - v_y)$. Such a coalition would form a *blocking* pair, which we want to rule out. Hence, our concept of stability implies

$$u_x + v_y \geq \Phi_{xy} \quad \forall x \in \mathcal{X}, \; y \in \mathcal{Y}, \tag{3.8}$$

for any pair x and y. Note that inequality (3.8) should hold for every pair (x, y), not only for the matched pair. Instead, the combination of inequality (3.8) and equality (3.7) implies that if x and y are matched under π (i.e., if $\pi_{xy} > 0$), then (3.8) should hold as an equality. We formalize this into the following definition.

DEFINITION 3.2. *An outcome (π, u, v) is called stable when*

   (i) $\pi \in \mathcal{M}(p, q)$; and
   (ii) *feasibility condition (3.7) holds; and*
   (iii) *no-blocking-pair condition (3.8) holds.*

Stability is closely connected to optimality, as implied by the following result.

PROPOSITION 3.3. *An outcome (π, u, v) is stable if and only if π is a solution to (3.3) and (u, v) is a solution to (3.4).*

PROOF. Assume that (π, u, v) is stable. Then condition (i) in definition 3.2 implies that π is feasible for the primal problem, and condition (iii) implies that (u, v) is feasible for the dual problem. Therefore condition (ii) implies that π is optimal for the primal and (u, v) is optimal for the dual. The converse is also straightforward. □

3.3 PURE ASSIGNMENTS

We have stated the optimal assignment problem in Monge–Kantorovich form, but we would like to return to the Monge problem, where every worker of type x is matched to jobs of the same type. For this problem to make sense, we have to assume that there is the same number of workers as jobs, denoted n ($n = N = M$), and there is one individual per type (i.e., $p_x = q_y = 1/n$). These assumptions will be maintained throughout this section. In this case, a matrix $\pi \in \mathcal{M}(P, Q)$ is written as $\pi = \Pi/n$, where Π satisfies

$$\sum_{y=1}^{n} \Pi_{xy} = 1 \quad \text{and} \quad \sum_{x=1}^{n} \Pi_{xy} = 1. \tag{3.9}$$

Matrices satisfying these conditions are called *doubly stochastic matrices*. Up to a scaling factor n, the value of the Monge–Kantorovich problem is therefore

$$\max_{\Pi \geq 0} \sum_{xy} \Pi_{xy} \Phi_{xy}$$

$$\text{s.t.} \sum_{y=1}^{n} \Pi_{xy} = 1 \quad \text{and} \quad \sum_{x=1}^{n} \Pi_{xy} = 1. \tag{3.10}$$

Then it makes sense to consider assignments such that each worker works for one and only one job, and such that each job employs one and only one worker. Such a map assigns a worker's index in $\{1, \ldots, n\}$ into the assigned job's index in $\{1, \ldots, n\}$; it is therefore an invertible map from $\{1, \ldots, n\}$ onto itself, which is called a *permutation*. Let $\mathfrak{S}_n$ denote the set of permutations of $\{1, \ldots, n\}$. If $\sigma \in \mathfrak{S}_n$ is a permutation, $\sigma(x)$ will denote the index of the job for which employee x works. The Monge problem is formulated as

$$\max_{\sigma \in \mathfrak{S}_n} \sum_{x=1}^{n} \Phi_{x\sigma(x)}, \tag{3.11}$$

which is an optimization problem over the set of permutations $\mathfrak{S}_n$, which is finite—hence problem (3.11) is a finite optimization problem. However, the set of permutations is a very large set, which has $n!$ elements. As a reminder, when $n = 20$, this number is approximately equal to $2 \cdot 10^{18}$. Therefore, an exhaustive search is hopeless beyond very small values of n.

Fortunately, it turns out that in the present setting, the Monge–Kantorovich problem has a pure solution, which implies that one may solve the Monge problem (3.11) by solving its linear programming relaxation (3.10). To each permutation $\sigma \in \mathfrak{S}_n$, we can associate the corresponding permutation matrix

$$\Pi^{\sigma}_{xy} := 1\{y = \sigma(x)\}, \tag{3.12}$$

which is equal to 1 if x and y are matched, and 0 otherwise. Obviously,

$$\sum_{x=1}^{n} \Phi_{x\sigma(x)} = \sum_{xy} \Pi^{\sigma}_{xy} \Phi_{xy},$$

hence we see that the value of (3.11) is less than or equal to the value of (3.10). But the values of the two problems actually coincide, as we are about to see in the next result.

THEOREM 3.4. *The value of the Kantorovich problem (3.10) coincides with the value of the Monge problem (3.11). Further, among the solutions to (3.10), there is a permutation matrix.*

PROOF. Because assignment matrices are doubly stochastic, the value of problem (3.11) is less than or equal to the value of problem (3.10). Let us show that they coincide. Let Π be a solution to problem (3.10), and assume that Π is not integral. Then, letting p be the number of entries of Π that are not integers, we will show that if $p > 1$, there exists a doubly stochastic matrix with at most $p - 1$ entries that are not integers, and that is still optimal. To see this, let (i_1, j_1) be a nonintegral entry of Π. It is easy to see that on row i_1 there is another nonintegral entry. Call it (i_2, j_2), with $i_2 = i_1$. By the same argument, there exists on column j_2 another nonintegral entry. Call it (i_3, j_3) with $j_3 = j_2$. Repeat this process until hitting a line or column already visited. This yields a cycle $(i_k, j_k), \ldots, (i_{k+l+1}, j_{k+l+1})$ with $i_{k+l+1} = i_k$ and $j_{k+l+1} = j_k$, and two adjacent entries on this cycle share alternately the same row or column. For $\epsilon \in \mathbb{R}$, consider the matrix Π^ϵ obtained by alternately adding and removing ϵ along this path, that is, by adding $(-1)^r \epsilon$ to entry (i_{k+r}, j_{k+r}) of Π for $0 \leq r \leq l$. We have $\Pi^\epsilon = \Pi + \epsilon \left(\Pi^1 - \Pi \right)$. Take the minimal and maximal values of ϵ such that all the entries in Π^ϵ remain between 0 and 1, and denote them respectively by $\underline{\epsilon} < 0$ and $\bar{\epsilon} > 0$. Then for $\epsilon \in \{\underline{\epsilon}, \bar{\epsilon}\}$, at least one entry in Π^ϵ along the path is 0, while all entries in Π along the path are nonzero. Outside the path, Π and Π^ϵ coincide. Thus if $\epsilon \in \{\underline{\epsilon}, \bar{\epsilon}\}$, the number of nonzero entries of Π^ϵ is strictly less than p. Further, Π^ϵ is clearly doubly stochastic. Finally, assume by contradiction that $\Pi^{\bar{\epsilon}}$ is suboptimal. Then $\sum_{xy} \Pi^{\bar{\epsilon}}_{xy} \Phi_{xy} < \sum_{xy} \Pi_{xy} \Phi_{xy}$; but as $\Pi^\epsilon = \Pi + \epsilon \left(\Pi^1 - \Pi \right)$, and as $\underline{\epsilon} < 0$ and $\bar{\epsilon} > 0$, this would imply that $\sum_{xy} \Pi^{\underline{\epsilon}}_{xy} \Phi_{xy} > \sum_{xy} \Pi_{xy} \Phi_{xy}$, a contradiction. Hence, $\Pi^{\bar{\epsilon}}$ is optimal (and so is $\Pi^{\underline{\epsilon}}$ by the same token). Hence, by induction, one can show that there is an optimal doubly stochastic matrix with zero noninteger entries. □

Note that the proof provided here is constructive: it provides an algorithm to obtain a pure optimal assignment from a general solution, which the reader is asked to implement in exercise 3.3. This result is in fact tightly connected to two important results in convex analysis: Choquet's theorem, which states that among the maximizers of a linear form over a convex compact set, there are extremal elements of that set, and the Birkhoff–von Neumann theorem, which states that the extremal elements of the set of doubly stochastic matrices are exactly the permutation matrices. Hence, there is a permutation matrix among the solutions to (3.10), and the value of that problem coincides with the value of (3.11).

3.4 COMPUTATION VIA LINEAR PROGRAMMING

We now turn to the issue of the practical computation of problem (3.3). This problem is arguably the most-studied problem in computer science, and dozens, if not hundreds, of algorithms exist, whose running time is polynomial in $\max(N, M)$, typically a power 3 of the latter. We won't have a full-length discussion of these algorithms here; we refer to [27] for exhaustive considerations. These algorithms have fast implementations in C, and there are R packages (such as `clue` [80]) which interface with them. Instead, we will show how to solve the problem with the help of a linear programming solver used as a black box; our challenge here will be to carefully set up the constraint matrix as a sparse matrix in order to let large-scale linear programming solvers[2] recognize and exploit the sparsity of the problem.

Let Π and Φ be the matrices with typical elements (π_{xy}) and (Φ_{xy}). We let p, q, u, v, and 1 be the column vectors with entries (p_x), (q_y), (u_x), (v_y), and 1, respectively. Problem (3.3) is rewritten using matrix algebra as

$$\max_{\Pi \geq 0} \mathrm{Tr}\left(\Pi'\Phi\right),$$

$$\Pi 1_M = p, \tag{3.13}$$

$$1_N'\Pi = q'.$$

This problem is clearly linear with respect to Π; however, note that in (3.13), the optimization variable Π is treated as a matrix, not as a vector. To solve this problem, we need to convert matrices into vectors; for instance, this can be done by stacking the columns of Π into a single column vector, which is typically the convention followed by most mathematical programming software. This operation is called *vectorization*, which we will denote

$$\mathrm{vec}\,(A),$$

which reshapes an $N \times M$ matrix into an $NM \times 1$ vector. In R, this is implemented by the instruction `c(A)`; in MATLAB, this is implemented by `reshape(A,[N,M])`. Hence, the matrix scalar product $\mathrm{Tr}\left(\Pi'\Phi\right)$ in the objective function will be replaced by the scalar product of the vectorized objects $\mathrm{vec}\,(\Phi)'\,\mathrm{vec}\,(\Pi)$.

Let us now look at the constraints. Clearly $\mathrm{vec}\,(\Pi 1_M)$ is a linear expression in $\mathrm{vec}\,(\Pi)$. The exact expression is provided by *Kronecker products*. If A is an $m \times p$ matrix and B an $n \times q$ matrix, then the Kronecker product $A \otimes B$ of A and B is an $mn \times pq$ matrix such that

$$\mathrm{vec}\left(BXA'\right) = (A \otimes B)\,\mathrm{vec}(X). \tag{3.14}$$

Thus, using Kronecker products, we can vectorize constraints that would otherwise be written as matrix products. In R, the Kronecker product of A and

---

[2] An optimization problem is called *large scale* when it does not require storing or operating on the full set of matrix entries; this is typically the case in a linear programming (LP) problem when the matrix defining the constraints is sparse. Nowadays, large-scale optimization can be readily implemented in R or MATLAB using an LP solver such as Gurobi and a standard toolbox for sparse matrices.

B is called `kronecker(A,B)`. In MATLAB, the corresponding instruction is `kron(A,B)`. The first constraint, which can be written as $I_N \Pi 1_M = p$, where I_N is the $N \times N$ identity matrix, vectorizes therefore as $\left(1'_M \otimes I_N\right) \text{vec}(\Pi) = \text{vec}(p)$, and similarly, the second constraint, which can be expressed as $1'_N \Pi I_M = q'$, vectorizes as $\left(I_M \otimes 1'_N\right) \text{vec}(\Pi) = \text{vec}\left(q'\right)$.

Note that the matrix $1'_M \otimes I_N$ is of size $N \times NM$, and the matrix $I_M \otimes 1'_N$ is of size $M \times NM$; hence the full matrix involved on the left-hand side of the constraints is of size $(N + M) \times NM$. In spite of its large size, this matrix is *sparse*, meaning that it has many zeros. Indeed, matrix I_N is sparse, thus $\left(1'_M \otimes I_N\right)$ is also sparse, and likewise for $I_M \otimes 1_N$. This remark is important for computational speed and efficiency considerations; however, we need to instruct the program to treat the matrices as sparse matrices. In order to do this, the identity matrix I_N is coded in R as `sparseMatrix(1:N,1:N)`, while in MATLAB it is coded as `Speye(N)`.

Setting $z = \text{vec}(\Pi)$, the linear programming problem then becomes

$$\max_{z \geq 0} \text{vec}(\Phi)' \, z$$
$$\text{s.t. } \left(1'_M \otimes I_N\right) z = \text{vec}(p), \tag{3.15}$$
$$\left(I_M \otimes 1'_N\right) z = \text{vec}\left(q'\right),$$

which is ready to be passed on to a linear programming solver such as Gurobi. An LP solver typically computes problems of the form

$$\max_{z \geq 0} c' z$$
$$\text{s.t. } Az = d, \tag{3.16}$$

with some flexibility to replace max by min, the equality constraints by inequality constraints, etc. In R, Gurobi is called to compute program (3.16) by `gurobi(list(A=A,obj=c,modelsense="max",rhs=d, sense="="))`.

PROGRAMMING EXAMPLE 3.1. The program `optimalAssignment.R` provides an R implementation of the optimal assignment problem using Gurobi. Note that Gurobi needs to be installed with an active licence (freely available to the academic community from www.gurobi.com). For comparison purposes, the code also computes the optimal assignment using the Hungarian algorithm via the R package `clue`. However, the latter does not provide the dual solutions u and v, and works only when $p_x = q_y = 1/n$.

To give an idea of the order of magnitude of the problems that can be solved in practice by linear programming, we run simulations on a laptop personal computer, middle of the range by 2015 standards. We simulate a matrix Φ

of uniform random numbers between 0 and 1, and we run the R program described above. With $n = 1000$, one gets the answer in a few seconds. With $n = 3000$ the result takes a bit more than a minute. With $n = 4000$, it takes around two minutes. With $n = 5000$, the program runs out of memory. Hence, problems of size n greater than a few thousand should be solved using other algorithms, or using a more powerful machine.

3.5 EXERCISES

Exercise 3.1 (M). ***Another proof of duality.*** *With the formalism developed in section 3.4, provide a proof of theorem 3.1 based purely on the matrix formulation of linear programming duality*

$$\max_{\substack{z \geq 0 \\ Az=d}} c'z = \min_{\substack{w \\ A'w \geq c}} d'w. \tag{3.17}$$

Exercise 3.2 (M). ***Matching with 0–1 costs.*** *Assume $c_{xy} \in \{0, 1\}$ and consider the problem*

$$V = \max_{u,v} \sum_x u_x - \sum_y v_y \tag{3.18}$$

$$\text{s.t. } u_x - v_y \leq c_{xy}.$$

   (i) *Show that the value of problem (3.18) is unchanged if one restricts the entries of u and v to be in $[0, 1]$.*
   (ii) *Show that if $(u_x, v_y) \in [0, 1]^2$, then $(u_x, v_y) = \int_0^1 (1\{t \leq u_x\}, 1\{t \leq v_y\})\, dt$.*
   (iii) *Assume (u, v) is feasible for problem (3.18). Show that for each $t \in [0, 1]$, the vectors (u^t, v^t), defined by $u_x^t = 1\{t \leq u_x\}$ and $v_y^t = 1\{t \leq v_y\}$, are feasible for problem (3.18).*
   (iv) *Show that the value of problem (3.18) is unchanged if one restricts the entries of u and v to be in $\{0, 1\}$.*

Exercise 3.3 (C). ***Purifying a matching.*** *Write a program that takes as an input a doubly stochastic matrix solution to (3.10) and returns a permutation matrix solution to (3.11). You can write a variant of programming example 3.1 for this purpose.*

Exercise 3.4 (C). ***A numerical experiment.*** *Generate 100 sample points from two bivariate Gaussian distributions of mean 0 and the variance–covariance matrix of your choice. Take $\theta \in [0, 2\pi]$, let $\Phi_\theta(x, y) = x_1 x_2 \cos \theta + y_1 y_2 \sin \theta$, and let π_θ be the solution to the optimal assignment problem between the sample* *points and surplus* Φ_θ. *Let*

$$C_\theta = \left(\mathbb{E}_{\pi_\theta}\left[X_1 X_2\right], \mathbb{E}_{\pi_\theta}\left[Y_1 Y_2\right]\right) \in \mathbb{R}^2$$

be the corresponding correlations. For each value of θ *on a grid compute* π_θ *(you can use the program in programming example 3.1 for this purpose), and plot* C_θ *in the plane. What do you notice?*

Exercise 3.5 (E). **Matching with singles.** *Consider a variant of problem (3.3) when the constraints are replaced by inequality constraints*

$$\sum_{y=1}^{M} \pi_{xy} \leq p_x \quad and \quad \sum_{x=1}^{N} \pi_{xy} \leq q_y.$$

Write the dual associated with the new problem, and provide the economic interpretation.

Exercise 3.6 (E). **Only types matter.** *Consider a population of individual workers and firms* $i \in \mathcal{I}$ *and* $j \in \mathcal{J}$. *The type of worker i (respectively, firm j) is* $x_i \in \mathcal{X}$ *(respectively,* $y_j \in \mathcal{Y}$*). Assume that the matching surplus between i and j depends on i and j only through their respective types, that is,* $\Phi_{ij} = \Phi_{x_i y_j}$. *The reservation utilities are all set to 0. Let* $\left(\pi_{ij}, u_i, v_j\right)$ *be a stable outcome. Show that* u_i *(respectively,* v_j*) depends only on* x_i *(respectively, on* y_j*), that is, there are vectors* (u_x) *and* $\left(v_y\right)$ *such that* $u_i = u_{x_i}$ *and* $v_j = v_{y_j}$.

3.6 REFERENCES AND NOTES

A classical textbook treatment for a large part of the material covered in this chapter is Roth and Sotomayor [125, chapter 8]; the main difference is that in our exposition, we do not allow for unassigned agents. The standard reference for section 3.1 is Koopmans and Beckmann [90]; see also Beckmann [13] and the treatise by Dantzig on linear programming [44, chapters 14, 15]. The reference for section 3.2 is Shapley and Shubik [138]. The results in section 3.3 originate from Birkhoff [19] and von Neumann [153]. A good textbook for the matrix algebra used in section 3.4 is Magnus and Neudecker [97]. A number of alternative algorithms to compute the assignment problem are discussed in the textbook by Burkard, Dell'Amico, and Martello [27]. The program in programming example 3.1 uses the Gurobi solver [74]. Exercise 3.2 is an ingredient of Strassen's theorem (see [141]), and our presentation follows Villani [148, chapter 1.4].

—4—

One-Dimensional Case

In this chapter, we consider the Monge–Kantorovich problem in the one-dimensional case, that is, when $\mathcal{X} = \mathcal{Y} = \mathbb{R}$. This implies that both workers and firms are characterized by scalar characteristics x and y. Although this seems restrictive, most of the applied literature in economics to this day has focused on this case, which already generates interesting insights. For instance, several authors have considered the assignment problem of firms to managers. The basic assumption of these models is that each manager has a measure of talent $x \in \mathbb{R}_+$, which captures the extra financial return generated by this individual. The relevant characteristic of each firm is market capitalization, denoted $y \in \mathbb{R}_+$. Then, the economic value generated by a manager of talent x running a firm of size y is the product of the extra rate of return and the firm market capitalization, that is, $\Phi(x, y) = xy$. It is assumed that there is the same number of managers as firms. The distribution of talent in the population of managers is P and the distribution of firm sizes is Q. Assume that a manager of talent x is assigned to a firm of size $y = T(x)$. The constraint on the assignment map T is that $T\#P = Q$, which means that each firm is run by a manager and each manager runs a firm. Then the total value created is

$$\mathbb{E}\left[\Phi\left(X, T(X)\right)\right] = \mathbb{E}\left[XT(X)\right].$$

Intuitively, it makes sense to expect that in the optimal solution, the most talented managers will run the largest firms. This is indeed the case under a natural assumption on Φ, called supermodularity, which is in particular satisfied by the multiplicative specification $\Phi(x, y) = xy$. As we shall see, the analysis of the primal Monge–Kantorovich problem will yield an optimal solution such that more talented managers run more sizeable firms, that is, $Y = T(X)$, where T is nondecreasing. An explicit characterization will be given to T. Further, we shall see that the dual Monge–Kantorovich problem will allow us to pin down the equilibrium wage curve $u(x)$.

4.1 COPULAS AND COMONOTONICITY

Throughout this chapter, the uniform distribution over $[0, 1]$, $\mathcal{U}([0, 1])$, will play an important role; we shall denote it by

$$\mathcal{U} := \mathcal{U}([0, 1]). \tag{4.1}$$

Let us recall that given a probability distribution P on the real line, one can define the quantile of that distribution, denoted F_P^{-1}, which is a map from $[0, 1]$ to $\mathbb{R}$, nondecreasing and continuous from the right, and such that if $U \sim \mathcal{U}$, then

$$F_P^{-1}(U) \sim P. \tag{4.2}$$

We denote by F_P^{-1} the generalized inverse of the c.d.f. of P, F_P. See appendix C for a definition and properties of generalized inverses. Note that when F_P is invertible, which is the case in particular when P has a positive density, then F_P^{-1} is the proper inverse of F_P, in which case $U = F_P(X)$. Note also that if $X \sim P$, then $F_P(X)$ has a uniform distribution if and only if P has no mass point.

Representation (4.2) extends to the case of bivariate distributions: for $\pi \in \mathcal{M}(P, Q)$, there exists a pair (U, V) of uniform random variables such that

$$\left(F_P^{-1}(U), F_Q^{-1}(V) \right) \sim \pi, \tag{4.3}$$

and, as seen in appendix C, the c.d.f. associated with the distribution of (U, V) is called the *copula* associated with distribution π. This simple fact has an important implication for the Monge–Kantorovich problem in the scalar case currently considered. Indeed, it is easy to see that the Monge–Kantorovich primal problem (2.6), combined with representation (4.3), can be rewritten as

$$\sup_{(U,V) \sim \lambda \in \mathcal{M}(\mathcal{U}, \mathcal{U})} \mathbb{E}_\lambda \left[\Phi \left(F_P^{-1}(U), F_Q^{-1}(V) \right) \right]. \tag{4.4}$$

Hence, in studying the Monge–Kantorovich problem in dimension one, we can consider the case $P = Q = \mathcal{U}$ without loss of generality. Another way of saying the same thing is that in dimension one, the Monge–Kantorovich problem is equivalent to an *extremal copula* problem.

In particular, we will be interested in a class of Monge–Kantorovich problems whose solutions are *comonotone*; these solutions are such that $U = V$ in representation (4.3). More precisely, we have the following definition.

DEFINITION 4.1. *A pair of random variables (X, Y) is comonotone if there is $U \sim \mathcal{U}$ such that $X = F_P^{-1}(U)$ and $Y = F_Q^{-1}(U)$. Equivalently, X and Y are said to exhibit* positive assortative matching (PAM).

Note that when the distribution of X is continuous, there is a much simpler equivalent statement of comonotonicity:

LEMMA 4.2. *If the distribution of X has no mass points, then X and Y are comonotone if and only if there exists a nondecreasing map T such that $Y = T(X)$. Moreover, one can choose $T(x) = F_Q^{-1}(F_P(x))$.*

PROOF. Consider $U \sim \mathcal{U}$ such that $X = F_P^{-1}(U)$ and $Y = F_Q^{-1}(U)$. If the distribution of X has no mass point, then $U = F_P(X)$. Hence, $Y = F_Q^{-1}(F_P(X))$. $\square$

The copula associated with a pair of comonotone random variables is the c.d.f. associated with (U, U), which is $F(u, v) = \min(u, v)$. This copula is called the *upper Fréchet–Hoeffding copula.*

4.2 SUPERMODULAR SURPLUS

Assume that Φ is *supermodular,* that is, for scalars x, x', y, and y',

$$\Phi\left(x \vee x', y \vee y'\right) + \Phi\left(x \wedge x', y \wedge y'\right) \geq \Phi(x, y) + \Phi\left(x', y'\right), \qquad (4.5)$$

where $x \vee x'$ and $x \wedge x'$ denote, respectively, the maximum and the minimum between scalars x and x'. When Φ is twice continuously differentiable (which we will assume from now on), this is equivalent to

$$\frac{\partial^2 \Phi(x, y)}{\partial x \partial y} \geq 0. \qquad (4.6)$$

Example 4.1. *The following examples of surplus functions are supermodular:*

(i) *Cobb–Douglas function:* $\Phi(x, y) = x^a y^b$ $(x, y \geq 0)$, *with $a, b \geq 0$*

(ii) *general multiplicative form:* $\Phi(x, y) = \zeta(x)\xi(y)$ *with ζ and ξ nondecreasing*

(iii) *Leontieff:* $\Phi(x, y) = \min(x, y)$

(iv) *C.E.S. function:* $\Phi(x, y) = (x^{-\rho} + y^{-\rho})^{-1/\rho}$, $\rho \geq 0$

(v) $\Phi(x, y) = \phi(x - y)$, *where ϕ is concave; in particular, $\Phi(x, y) = -|x - y|^p$, $p \geq 1$ or $\Phi(x, y) = -(x - y - k)^+$*

(vi) $\Phi(x, y) = \phi(x + y)$, *where ϕ is convex*

To understand the economic content of the supermodularity condition, assume that there are two types of workers: low type $\underline{x}$ and high type $\bar{x}$. Assume that similarly, there are two types of firms, the low type $\underline{y}$ and high type $\bar{y}$. An equivalent restatement of condition (4.5) is

$$\bar{x} \geq \underline{x} \text{ and } \bar{y} \geq \underline{y} \quad \text{implies} \quad \Phi(\bar{x}, \bar{y}) + \Phi(\underline{x}, \underline{y}) \geq \Phi(\bar{x}, \underline{y}) + \Phi(\underline{x}, \bar{y}), \qquad (4.7)$$

which asserts that the total output created is higher if the high types match together and the low types match together (assortative matching) rather than if mixed high/low pairs are formed.

The main result in this chapter asserts that when Φ is supermodular, it is optimal to match the higher types with the higher types, and the lower types with the lower types.

THEOREM 4.3.   (i) *Assume that Φ is supermodular. Then the primal of the Monge–Kantorovich problem*

$$\sup_{\pi \in \mathcal{M}(P,Q)} \mathbb{E}_\pi \left[\Phi(X, Y) \right] \tag{4.8}$$

has a comonotone solution.

(ii) *Conversely, if problem (4.8) has a comonotone solution for any choice of probability distributions P and Q on the real line, then Φ is supermodular.*

(iii) *If, in addition, P has no mass points, then there is an optimal assignment which is pure and satisfies $Y = T(X)$, where*

$$T(x) = F_Q^{-1} \circ F_P(x). \tag{4.9}$$

The proof of part (i) is based on the following lemma.

LEMMA 4.4. *Let Z_1 and Z_2 be two Bernoulli random variables of respective success probabilities p_1 and p_2. Then $\mathbb{E}\left[Z_1 Z_2\right] \leq \min\left(p_1, p_2\right)$.*

PROOF. As $Z_2 \leq 1$, then $\mathbb{E}\left[Z_1 Z_2\right] \leq \mathbb{E}\left[Z_1\right] = p_1$. Similarly $\mathbb{E}\left[Z_1 Z_2\right] \leq \mathbb{E}\left[Z_2\right] = p_2$. Thus, $\mathbb{E}\left[Z_1 Z_2\right] \leq \min\left(p_1, p_2\right)$. $\square$

We are now ready to sketch the proof of theorem 4.3.

SKETCH OF PROOF OF THEOREM 4.3.   (i) Take $U \sim \mathcal{U}$, and $X = F_P^{-1}(U)$ and $Y = F_Q^{-1}(U)$. By (4.2), $X \sim P$ and $Y \sim Q$ and (X, Y) is comonotone by definition. The proof is in three steps.

Step 1. For $a, b \in \mathbb{R}$, consider surplus function $\phi_{ab}(x, y) :=$ $1\{x \geq a\} 1\{y \geq b\}$, and let $Z_1 = 1\{X \geq a\}$ and $Z_2 = 1\{Y \geq b\}$. Here, Z_1 and Z_2 are two Bernoulli random variables of respective success probabilities $p_1 = 1 - F_P(a)$ and $p_2 = 1 - F_Q(b)$; thus $\mathbb{E}\left[Z_1 Z_2\right] \leq \min\left(p_1, p_2\right)$, but a straightforward calculation shows that the inequality actually holds as an equality. Hence (X, Y), which is comonotone, is optimal for each surplus function ϕ_{ab}.

Step 2. Assume $\mathcal{X} = [\underline{x}, \bar{x}]$ and $\mathcal{Y} = [\underline{y}, \bar{y}]$ are compact intervals, and let Φ be a supermodular function defined on $\mathcal{X} \times \mathcal{Y}$. Define

$$F(x, y) = \frac{\Phi(x, y) - \Phi(\underline{x}, y) - \Phi(x, \underline{y}) + \Phi(\underline{x}, \underline{y})}{\Phi(\bar{x}, \bar{y}) - \Phi(\underline{x}, \bar{y}) - \Phi(\bar{x}, \underline{y}) + \Phi(\underline{x}, \underline{y})}$$

is a c.d.f. associated to a probability measure ζ, and hence $F(x, y) = \iint \phi_{ab}(x, y) \, d\zeta(a, b)$. As a result, if $\pi \in \mathcal{M}(p, q)$ is the distribution of (X, Y) where X and Y are comonotone, then

$$\int F(x, y) \, d\pi(x, y) \geq \int F(x, y) \, d\tilde{\pi}(x, y)$$

for every $\tilde{\pi} \in \mathcal{M}(p, q)$. But as F is of the form $F(x, y) = K\Phi(x, y) + f(x) + g(y) + c$ with $K > 0$, and because $\int \{f(x) + g(y) + c\} \, d\pi(x, y) = \int \{f(x) + g(y) + c\} \, d\tilde{\pi}(x, y)$ for every $\tilde{\pi} \in \mathcal{M}(p, q)$, it results that

$$\int \Phi(x, y) \, d\pi(x, y) \geq \int \Phi(x, y) \, d\tilde{\pi}(x, y) \quad \forall \tilde{\pi} \in \mathcal{M}(p, q),$$

thus (X, Y) is optimal for surplus Φ, which completes step 2.

Step 3. When $\mathcal{X}$ and $\mathcal{Y}$ are the real line, the result still holds by approximation.

(ii) The converse follows by taking for P the discrete probability with two mass points $\underline{x}$ and $\bar{x}$ with probability $\frac{1}{2}$ each, and for Q the discrete probability with two mass points $\underline{y}$ and $\bar{y}$ also each with probability $\frac{1}{2}$. Then if (4.8) has a solution such that $X = F_P^{-1}(U)$ and $Y = F_Q^{-1}(U)$, for $U \sim \mathcal{U}([0, 1])$, it follows that condition (4.7) holds.

(iii) follows from (i) and lemma 4.2. □

Let us now discuss two important consequences of theorem 4.3. First, let us see what it implies in the discrete case. Assume P and Q are empirical distributions, namely, $\mathcal{X} = \mathcal{Y} = \{1, \ldots, n\}$ and $P = Q$ are discrete uniform distributions with mass points $p_x = q_y = 1/n$ for all x and y. Assume that Φ is supermodular. Then the identity map leads to a comonotone assignment, which is optimal by virtue of theorem 4.3(i). The next result follows.

COROLLARY 4.5. *For every permutation* $\sigma \in \mathfrak{S}_N$,

$$\sum_{x=1}^{n} \Phi_{xx} \geq \sum_{x=1}^{n} \Phi_{x\sigma(x)}. \tag{4.10}$$

This inequality is known as the Hardy–Littlewood–Pólya–Lorentz inequality. In particular, applied to $\Phi_{xy} = -|a_x - b_y|^p$, $p \geq 1$, the corollary implies that if for each $x \geq x'$, we have $a_x \geq a_{x'}$ and $b_x \geq b_{x'}$, then

$$\sum_{x=1}^{n} |a_x - b_x|^p \leq \sum_{x=1}^{n} |a_x - b_{\sigma(x)}|^p,$$

which implies that if a_x is nondecreasing in x, then the increasing rearrangement of b is weakly closer to a than b for any L^p-norm, $p \geq 1$.

This property has been used in economic theory for the comparison of risks, and in econometrics for shape-constrained estimation; see the references at the end of the chapter.

As a second consequence we show that, unsurprisingly, two comonotone variables are positively correlated.

COROLLARY 4.6. *Assume that X and Y are comonotone. Then the covariance between X and Y is nonnegative.*

PROOF. Let $\bar{x} = \mathbb{E}_P[X]$ and $\bar{y} = \mathbb{E}_Q[Y]$. Since X and Y are comonotone, they are optimal for the Monge–Kantorovich problem associated with $\Phi(x, y) = (x - \bar{x})(y - \bar{y})$, which is supermodular. Consider $(\tilde{X}, \tilde{Y})$, an independent coupling of P and Q. By optimality of (X, Y), then $\mathbb{E}[\Phi(X, Y)] \geq \mathbb{E}[\Phi(\tilde{X}, \tilde{Y})]$, which is to say that $\text{cov}(X, Y) \geq 0$. $\square$

In applied work, positive covariance is sometimes used as a testable implication of comonotonicity. However, two caveats are in order. First, the converse to corollary 4.6 is false: it is easy to see that positive covariance between X and Y does not imply comonotonicity between them. Second, the fact that X and Y are comonotone does not imply that their correlation coefficient is 1. Indeed, by the Cauchy–Schwarz inequality, the correlation coefficient between X and Y is 1 if and only if these random variables are linearly dependent, that is, $Y = aX + b$, with $a > 0$. In general this is far stronger than comonotonicity, which essentially expresses that $Y = T(X)$ with T nondecreasing.

Note that the assumptions made in theorem 4.3 do not guarantee that all the optimal assignments are comonotone. Indeed, the trivial example where $\Phi(x, y) = 0$ for every x and y provides an example of a supermodular surplus function, for which any assignment is optimal. For this reason, we provide a strengthening of the previous result, which ensures uniqueness. We will assume that Φ is strictly supermodular, that is, if both $\bar{x} \geq \underline{x}$ and $\bar{y} \geq \underline{y}$ hold, and one of these inequalities is strict, then $\Phi(\bar{x}, \bar{y}) + \Phi(\underline{x}, \underline{y}) > \Phi(\bar{x}, \underline{y}) + \Phi(\underline{x}, \bar{y})$.

THEOREM 4.7. *Assume that Φ is strictly supermodular, and P has no mass point. Then the primal Monge–Kantorovich problem (4.8) has a unique optimal assignment, and this assignment is characterized by $Y = T(X)$, where T is given by (4.9).*

The proof of this result is omitted, and the reader is referred, for example, to [26].

Historically, many economic applications have relied on scalar types and supermodular surpluses, despite the strength of these assumptions. The following chapters will show how to dispense with these assumptions.

4.3 THE WAGE EQUATION

We now turn our focus to the dual of the Monge–Kantorovich problem. As we shall see, the problem has a solution in this case, which can be characterized in closed form.

THEOREM 4.8.   *(i) Assume that Φ is supermodular and continuously differentiable with respect to its first variable. Assume that P has no mass point. Then the dual of the Monge–Kantorovich problem,*

$$\inf \mathbb{E}_P\left[u(X)\right] + \mathbb{E}_Q\left[v(Y)\right] \\ \text{s.t. } u(x) + v(y) \geq \Phi(x, y), \tag{4.11}$$

has a solution (u, v). Further, u is such that

$$u'(x) = \frac{\partial \Phi}{\partial x}(x, T(x)), \tag{4.12}$$

where T is given by (4.9). Hence, u is determined up to a constant c by

$$u(x) = c + \int_{x_0}^{x} \frac{\partial \Phi}{\partial x}(t, T(t))\, dt. \tag{4.13}$$

(ii) Assume further that Q has no mass point, and that Φ is also continuously differentiable with respect to its second variable. Then v is given by

$$v(y) = c' + \int_{T(x_0)}^{y} \frac{\partial \Phi}{\partial y}\left(T^{-1}(z), z\right) dz, \tag{4.14}$$

where c and c' are related by $c + c' = \Phi(x_0, T(x_0))$.

The idea of the proof is simple: If a solution (u, v) to the dual problem exists, then the firm's problem is

$$v(y) = \max_{x \in \mathcal{X}} \{\Phi(x, y) - u(x)\};$$

thus (assuming T is invertible), firm y will pick the optimal employee $x = T^{-1}(y)$. By first-order conditions, $u'(x) = \partial \Phi / \partial x\,(x, T(x))$, which is integrated into (4.13). A similar logic yields equation (4.14), where the integration constant is set by equality $u(x_0) + v(T(x_0)) = \Phi(x_0, T(x_0))$.

PROOF OF THEOREM 4.8. Assume that a solution (u, v) to the dual exists. As P has no mass point, theorem 4.3 implies that the optimal coupling (X, Y) satisfies $Y = F_Q^{-1}(F_P(X))$. Worker x and firm y match if and only if $u(x) + v(y) = \Phi(x, y)$, hence if $v(y) = \sup_x (\Phi(x, y) - u(x))$. By first-order conditions, this implies (4.12).

Now let $u(x) = c + \int_0^x (\partial \Phi(z, T(z))/\partial x)\, dz$, and let $v(y) = \Phi\left(T^{-1}(y), y\right) - u\left(T^{-1}(y)\right)$. One has

$$u(x) + v(y) = u(x) - u\left(T^{-1}(y)\right) + \Phi\left(T^{-1}(y), y\right).$$

Assume $x \geq T^{-1}(y)$ (the other case is treated in a similar fashion). Then

$$u(x) + v(y) = \int_{T^{-1}(y)}^x \frac{\partial \Phi}{\partial x}(z, T(z))\, dz + \Phi\left(T^{-1}(y), y\right), \tag{4.15}$$

but

$$\Phi(x, y) - \Phi\left(T^{-1}(y), y\right) = \int_{T^{-1}(y)}^x \partial_x \Phi(z, y)\, dz$$

and $z \geq T^{-1}(y)$ implies $y \leq T(z)$, which implies by supermodularity of Φ that $\partial_x \Phi(z, y) \leq \partial_x \Phi(z, T(y))$. Hence,

$$\Phi(x, y) - \Phi\left(T^{-1}(y), y\right) \leq \int_{T^{-1}(y)}^x \partial_x \Phi(z, T(y))\, dz.$$

Combining with (4.15) yields $u(x) + v(y) \geq \Phi(x, y)$, thus the pair (u, v) is feasible for the dual. Finally, by construction,

$$\int u(x)\, dP(x) + \int v(y)\, dQ(y) = \int \Phi\left(T^{-1}(y), y\right)\, dQ(y),$$

which is the value of the primal. Hence (u, v) is optimal. $\square$

It is important to distinguish between $\partial \Phi(x, T(x))/\partial x$, which is the partial derivative of $\Phi(x, y)$ applied at $(x, y) = (x, T(x))$, and $d\Phi(x, T(x))/dx$, which is the total derivative of $\Phi(x, T(x))$ with respect to x. One has

$$\frac{d\Phi(x, T(x))}{dx} = \frac{\partial \Phi(x, T(x))}{\partial x} + \frac{\partial \Phi(x, T(x))}{\partial y} T'(x),$$

which has an interesting interpretation in terms of differential rent. The total derivative $d\Phi(x, T(x))/dx$ is the marginal increase in value between a firm run by a manager of talent x and a firm run by a manager of talent $x + dx$. This differential value is split between the manager's differential rent and the firm's differential rent. The manager's differential rent, which is the marginal wage increase, is $\partial \Phi(x, T(x))/\partial x = u'(x)$, while the firm's differential rent is the marginal increase in its profit, that is, $(\partial \Phi(x, T(x))/\partial y)\, T'(x) = dv(T(x))/dx$. This discussion highlights the nature of the assignment model, which is a model of imperfect competition. In this model, managers compete against each other, and likewise on the other side of the market. Managers are imperfect substitutes for each other, in the sense that for a given firm y and two distinct managers x and x', $\Phi(x, y) \neq \Phi(x', y)$. This imperfect competition is the source of the rent, as made apparent by formula (4.12). Of course, as before, the outside options are left unspecified, which explains why the wage curve in formula (4.13) is specified only up to a constant c to be determined.

We close this section with an example where the sharing of the matching surplus has a particularly simple form.

Example 4.2. *Assume that the surplus has a Cobb–Douglas form*

$$\Phi(x, y) = x^a y^b, \tag{4.16}$$

where $a, b > 0$ are output elasticities of managerial talent and firm size respectively. Assume that both P and Q are the uniform distribution $\mathcal{U}([0, 1])$. Then the optimal assignment is $T(x) = x$. Further, $u'(x) = ax^{a-1} T(x)^b = ax^{a+b-1}$, so

$$u(x) = c + \frac{a}{a+b} x^{a+b} \quad and \quad v(y) = -c + \frac{b}{a+b} y^{a+b}. \tag{4.17}$$

These formulas have a very simple interpretation. In this setting, a manager of talent x matches with a firm of size x. The economic value generated is then $\Phi(x, x) = x^{a+b}$. This value is split between managers and firms in proportion to the output elasticities a and b.

4.4 NUMERICAL COMPUTATION

The numerical implementation of the types of models seen in this chapter poses no particular challenge. Indeed, the optimal assignment $T(x)$ is given by

$$T(x) = F_Q^{-1} (F_P(x))$$

and both the c.d.f. and the quantile functions are usually routinely computed in most mathematical programming software. In R in particular, these functions are available for almost any known parametric distribution. For instance, in the case of the normal distribution, these functions are implemented respectively by pnorm and qnorm. For the gamma distribution, the functions are pgamma and qgamma etc. Turning to the determination of the wage curve $u(x)$, this is performed using numerical integration of

$$c + \int_{x_0}^x \frac{\partial \Phi}{\partial x} (t, T(t)) \, dt.$$

Indeed, this integral is not available in closed form, except in some simple cases. Here again, there is no major obstacle to fear, as numerical integration over the real line is a simple task. In R, this is performed using a call to `integrate`.

The following programming example provides a demonstration of such computations at work.

PROGRAMMING EXAMPLE 4.1. The program `PE-PositiveAssort-ativeMatching.R` numerically computes $T(x)$, $u(x)$, and $v(y)$, based on the specifications of `cdf_P`, `quantile_Q`, `dPhi_dx`, and `dPhi_dy`.

4.5 EXERCISES

Exercise 4.1 (M). *Multivariate c.d.f. Let $X \sim P$ be an $\mathbb{R}^d$-valued random vector, $d \geq 2$. Let F_P be its cumulative distribution function, that is, $F_P(x) = \Pr(X \leq x)$, where $\leq$ is the componentwise order of $\mathbb{R}^d$. Give an example to show that the distribution of $F_P(X)$ differs from $\mathcal{U}([0,1])$ in general.*

Exercise 4.2 (M). *The RUSC model. We study a Monge–Kantorovich problem when the distribution P of X is the uniform distribution over $[0,1]$, and Q is a finitely supported distribution with $M + 1$ atom points y_j, $0 \leq j \leq M$, and the mass of y_j is q_j. Without loss of generality one may assume $y_j < y_{j'}$ for $j < j'$, and $\Phi(x, y) = xy$. The social surplus is given by the value of the Monge–Kantorovich problem, that is,*

$$\max_{\pi \in M(P,Q)} \mathbb{E}_{\pi}[XY].$$

This model is called the random uniform scalar coefficient (RUSC) model in Galichon and Salanié [69].

   (i) *Show that this value is $\frac{1}{2} \sum_{k,j} \max(y_k, y_j) \, q_j q_k$.*

   (ii) *Show that the solution $(u(x), v_j)$ of the dual Kantorovich problem is given by $u(x) = \max_{j=0,\ldots,M} \{xy_j - v_j\}$, $v_j = \sum_{k \neq 0} A_{jk} q_k + b_j$, where $A_{jk} = \max(y_k, y_j) - \max(y_k, y_0) - \max(y_j, y_0) + y_0$ and $b_k = \max(y_k, y_0) - y_0$.*

Exercise 4.3 (C). *Simulation versus closed form. Let $\Phi(x, y) = xy$. Assume the distribution P is $\mathcal{U}([0,1])$, and the distribution Q is a uniform distribution over $\{0.1, 0.3, 0.6, 1\}$, whose points have probability 0.25 each.*

   (i) *Draw a random sample of $N = 10{,}000$ points from P and compute the optimal assignment between samples $\{x_1, \ldots, x_N\}$ and $\{y_1, \ldots, y_4\}$, as well as the prices v_j, $(j = 1, \ldots, 4)$.*

   (ii) *Using the results of exercise 4.2, compare the prices obtained in (i) and the prices v_j obtained in closed form.*

Exercise 4.4 (C). *Gini index.* Write down code to compute the Gini index (look it up if needed) associated with the wage distribution $u(X)$, when $\Phi(x, y) = x^a y^b$, P is the uniform distribution, and Q is a gamma distribution. It is assumed that $u(0) = 0$.

Exercise 4.5 (E). *Capital versus labor.* The production function is assumed to satisfy $\Phi \geq 0$, $\partial_x \Phi \geq 0$, $\partial_y \Phi \geq 0$, and $\partial^2_{xy} \Phi \geq 0$. Assume that workers are in excess supply, and that the reservation wage is 0, so that the lowest-paid worker receives wage 0.

   (i) What is the effect of a technological shock $f(x) \geq 0$ in the productivity, so that production function $\Phi(x, y)$ is replaced by $\Phi(x, y) + f(x)$? Does this change necessarily increase the salary of the workers?
   (ii) What is the effect of a homogenous technological shock $g(y) \geq 0$ in the productivity, so that the production function $\Phi(x, y)$ is replaced by $\Phi(x, y) + g(y)$?

Exercise 4.6 (E). *Fiscal gain from marriage.* Assume that individuals marry only for fiscal purposes. Assume that the tax levied on a single individual with income x is $\tau(x)$, where the tax schedule τ is convex, while the tax levied on a household where x and y are the couple's incomes is $2\tau\left(\frac{1}{2}(x + y)\right)$. Write down the fiscal gain from marriage $\Phi(x, y)$. What are the properties of the optimal matching?

4.6 REFERENCES AND NOTES

Condition (4.6) has been given (up to variants) many names: *increasing differences, Spence–Mirrlees condition, positive assortative matching,* and *single crossing* in economics; *supermodularity* and *twist condition* in mathematics; *Monge condition* in computer science. These are not equivalent, but they all convey a similar idea. Theorem 4.3 is associated in economics with the name of Gary Becker and his pioneering analysis of the marriage market [11, 12]. In mathematics, this result is known as the Lorentz theorem [96], building on the work of Hardy, Littlewood, and Pólya [76]. Important references for theorem 4.8 are a series of papers by Michael Sattinger [133, 134, 135]. See recent applications to managerial compensation in Tervio [142] and in Gabaix and Landier [58]. Corollary 4.5 has been applied in economic theory in connection with stochastic orders (see [126]) and efficient risk-sharing (see [144, 93]); in econometrics, see, for example, [35, 36]. Corollary 4.5 is also a key tool in the study of stochastic orders; see the books [110, 137]. The reference for exercise 4.2 is the model called *random uniform scalar coefficient* (RUSC) in Galichon and Salanié [69].

5

Power Diagrams

In this chapter, we consider another instance of the Monge–Kantorovich problem, where the distribution P of X is continuous, the distribution Q of Y is discrete with a finite number of atom points, and the surplus is the scalar product, that is, $\Phi(x, y) = x' y$. We will relate this setting to Hotelling's celebrated location model in the next paragraph. This framework has similarities to the previous chapters, which it extends in several respects. As in chapter 3, it assumes that distribution Q is discrete; however, in contrast with that chapter, it assumes that distribution P is continuous. The scalar-product surplus extends the one-dimensional product surplus $\Phi(x, y) = xy$ which was one of the leading examples of supermodular surpluses in chapter 4; however, the present framework now allows x and y to be vectors of dimension greater than 1.

Under these assumptions, we shall see that the dual Monge–Kantorovich problem has a solution, and that the Monge problem has a solution. The existence of a solution (u, v) to the dual Monge–Kantorovich problem will come from the fact that the latter can be reformulated as a convex optimization problem in finite dimensions; indeed, the candidate solutions to the dual problem can be parameterized by a (finite-dimensional) vector consisting of the values taken by v on the (finite) support of Q. Further, we shall see that $u(x)$ is a convex and piecewise affine function, and these simple considerations will yield powerful insights into the solution to the problem. We will show that the Monge–Kantorovich theorem in this context is closely connected to several very useful tools in spatial analysis and computational geometry, namely, *Voronoi tessellations* and *power diagrams*. As we shall see, the set of x values that are assigned under a Monge solution map to a given value of y form a partition of the space $\mathcal{X}$ into a power diagram, which is a slight generalization of a Voronoi tessellation. Readers who are not familiar with these concepts should not worry: everything will be defined in due course.

5.1 HOTELLING'S LOCATION MODEL

Consider a city, where location is represented by $x \in \mathcal{X}$; $\mathcal{X}$ will be a subset of $\mathbb{R}^d$, $d = 2$, which will be assumed to be convex and polyhedral. Actually, $d = 2$

is not required except for expositional purposes; our presentation holds for any $d \geq 1$. Assume that there are M fountains, located at points $y_j \in \mathbb{R}^d$, $1 \leq j \leq M$. The location of inhabitants is distributed according to a continuous distribution P whose support is included in $\mathcal{X}$. It is assumed that the total mass of inhabitants and fountains is normalized to 1, which means that the total supply equals the total demand. Assume that an inhabitant located at x has a transportation cost associated with using fountain j that is proportional to the square of the distance to this fountain, which results in a surplus associated with using fountain j equal to

$$\tilde{\Phi}(x, y_j) := -\tfrac{1}{2} \left| x - y_j \right|^2, \tag{5.1}$$

which expresses a transportation cost equal to the square of the Euclidean distance between the inhabitant and the fountain. The choice of the squared distance may seem arbitrary, and another natural choice would be the distance itself, or a more general power of the distance. However, as we shall see in the next paragraph, the square distance leads to important algebraic simplification in the analysis, and a nice geometrical structure in the optimal assignment. The more general theory will be presented in chapter 7.

Without any usage fee, each inhabitant would choose the fountain nearest to their own location. We enhance the model by introducing prices that are charged by fountains, and we assume that utilities are quasi-linear in money. Let $\tilde{v}_j$ be the price charged by fountain j. The utility of the consumer at location x is therefore $\tilde{\Phi}(x, y_j) - \tilde{v}_j$, and the indirect surplus of the consumer at x is given by

$$\tilde{u}(x) = \max_{j \in \{1,\dots,M\}} \left\{ \tilde{\Phi}\left(x, y_j\right) - \tilde{v}_j \right\}, \tag{5.2}$$

which expresses the fact that consumers look for fountains that offer the best trade-off between distance (as captured by $\tilde{\Phi}$) and price charged (as captured by $\tilde{v}$).

5.1.1 A Reformulation

Before we proceed, we will slightly simplify the mathematical formulation of the problem by showing that one can replace without loss of generality the quadratic surplus $\tilde{\Phi}(x, y) = -\tfrac{1}{2} \left| x - y \right|^2$ by the scalar-product surplus

$$\Phi(x, y) := x'y. \tag{5.3}$$

Indeed, note that $\tilde{\Phi}(x, y) = \Phi(x, y) - \tfrac{1}{2}|x|^2 - \tfrac{1}{2}|y|^2$, and introducing the *reduced indirect surplus* $u(x)$ and the *reduced prices* v_j as

$$u(x) = \tilde{u}(x) + \tfrac{1}{2}|x|^2 \quad \text{and} \quad v_j = \tilde{v}_j + \tfrac{1}{2} \left| y_j \right|^2, \tag{5.4}$$

one immediately sees that $\tilde{u}(x) + \tilde{v}_j \geq \tilde{\Phi}\left(x, y_j\right)$ if and only if $u(x) + v_j \geq \Phi\left(x, y_j\right)$. It follows that the consumer at location x chooses fountain j that maximizes

$$u(x) = \max_{j \in \{1,\dots,M\}} \left\{ \Phi\left(x, y_j\right) - v_j \right\}. \tag{5.5}$$

Hence the problem can be reexpressed so that the surplus of consumer x at fountain j is simply $x'y_j - v_j$. It is clear from (5.5) that (unlike $\tilde{u}$), the reduced surplus u is a piecewise affine and convex function from $\mathbb{R}^d$ to $\mathbb{R}$. The connection with convex and piecewise affine functions is the reason for reformulating the problem as we did.

5.1.2 Geometry

The demand set of fountain j is the set of consumers who prefer using that fountain over using any other fountain, that is,

$$\mathcal{X}_j^v := \left\{ x \in \mathcal{X} : \tilde{\Phi}\left(x, y_j\right) - \tilde{v}_j \geq \tilde{\Phi}(x, y_k) - \tilde{v}_k \ \forall k \right\}. \tag{5.6}$$

When there is no ambiguity, we will drop the dependence on v and simply denote by $\mathcal{X}_j$ the set of consumers demanding fountain j. Note that we don't really worry whether the inequality holds weakly or strictly, as this occurs on zero-probability events. Up to these boundaries, the $\mathcal{X}_j$ form a partition of $\mathcal{X}$ which is called a *power diagram*. This graph has interesting properties. First, each $\mathcal{X}_j$ is a convex polyhedron. Indeed, $x \in \mathcal{X}_j$ is equivalent to

$$x'\left(y_j - y_k\right) \geq v_j - v_k \quad \forall k. \tag{5.7}$$

As a result, it becomes apparent that the intersection of $\mathcal{X}_j$ and the $\mathcal{X}_k$ is supported by the hyperplane of equation $\{x : x'(y_j - y_k) + v_k - v_j = 0\}$. Furthermore, one sees that the set $\mathcal{X}_j$ weakly increases when v_k $(k \neq j)$ increases, and strictly decreases when v_j decreases.

If fountains do not charge any fee, that is, if $\tilde{v}_j = 0$, or equivalently if $v_j = \frac{1}{2}|y_j|^2$, then $\mathcal{X}_j^0$ is the set of consumers who are closer to fountain j than to any other fountain. The cells $\mathcal{X}_j^0$ form a partition of $\mathcal{X}$ called a *Voronoi tessellation*, which is a very particular case of a power diagram; see figure 5.1. Voronoi diagrams have the property that fountain j belongs to cell $\mathcal{X}_j^0$; when $\tilde{v} \neq 0$, this property may no longer hold for more general power diagrams.

5.1.3 Analytical Considerations

The demand for fountain j is given by $P(\mathcal{X}_j) = \Pr(X \in \mathcal{X}_j)$, which is the mass of consumers who prefer fountain j over the others. We would like an analytic determination of these quantities. Note that

$$\mathcal{X}_j = \arg\max_{x \in \mathcal{X}} \left\{ x' y_j - u(x) \right\}, \tag{5.8}$$

but by the envelope theorem, whenever u is differentiable at x, then $x \in \mathcal{X}_j$ if and only if $\nabla u(x) = y_j$. Hence, in the sequel, we will work with surplus $x'y$ instead of $-\frac{1}{2}|x - y|^2$. Therefore

$$\mathcal{X}_j := \nabla u^{-1}(\{y_j\}). \tag{5.9}$$

We can provide an elegant determination of $P\left(\mathcal{X}_j^v\right) = \Pr_P\left(X \in \mathcal{X}_j^v\right)$, the total demand for fountain j. Introduce the social welfare of consumers for the reduced price vector v as the sum of the reduced surplus across the population, that is,

$$W(v) := \mathbb{E}_P\left[\max_{j \in \{1,\dots,M\}} \left\{ X' y_j - v_j \right\}\right]. \tag{5.10}$$

We have $\partial W(v)/\partial v_k = -\mathbb{E}_P[1\{x' y_k - v_k \geq x' y_j - v_j \ \forall j\}] = -P(\mathcal{X}_k^v)$. Thus, we summarize our findings in the following theorem.

THEOREM 5.1. *The set of individuals demanding fountain j is characterized by $\mathcal{X}_j = \{x \in \mathcal{X} : \nabla u(x) = y_j\}$. The total demand for fountain j is given by*

$$P\left(\mathcal{X}_k^v\right) = -\frac{\partial W(v)}{\partial v_k}, \tag{5.11}$$

where W is defined by (5.10).

What we have done so far is to treat prices as exogenous, and deduce demand from prices. In the sequel we shall assume that there is a fixed capacity for each fountain, and study the price adjustment mechanism.

5.2 CAPACITY CONSTRAINTS

Let us now introduce capacity constraints for each fountain. Let q_j be the capacity of fountain j, and assume that the total capacity of the M fountains j is 1, that is, $\sum_{j=1}^{M} q_j = 1$. Thus, while the total demand can be served in the aggregate, we need to ensure that inhabitants are properly assigned to fountains so that each fountain's capacity exactly meets its demand. Indeed, there is no guarantee that the demand for fountain j, which is $P(\mathcal{X}_j^v)$, should

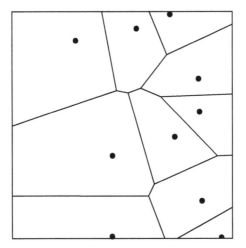

Figure 5.1: A Voronoi diagram with 10 cells. Each point x in the square is sent to the closest y_j; as a result, the areas of the cells are uneven. Here, there is a greater concentration of fountains in the right-hand part of the graph.

be equal to the capacity of fountain j, which is equal to q_j. Instead, there may be regions where there is a higher concentration of fountains, and regions where there is a lower concentration of them. Hence some fountains will be oversupplied, while some other fountains will be undersupplied, as shown in figure 5.1.

Classically, there are two ways to impose market clearing: either in a coercive way by the intervention of a central planner, or in a decentralized way through a system of prices which adjust at equilibrium. Without prices, a central planner is needed.

If coercive power is available, the central planner can decide on the assignment of consumers to fountains, in order to ensure that the market clears. This will be done by assigning to each inhabitant x a fountain $T(x) \in \{y_1, \ldots, y_M\}$, in a such way that each fountain j is used to its full capacity, that is,

$$P\left(T(X) = y_j\right) = q_j \quad \forall j \in \{1, \ldots, M\}. \tag{5.12}$$

Such a configuration is exemplified in figure 5.2, where the y_j are the same as in figure 5.1, but where consumers are assigned so that the balance equations (5.12) are satisfied. While there are many possible choices of T, the central planner may want to choose to optimize some welfare criterion under the balance constraints (5.12). One natural welfare criterion is to maximize the

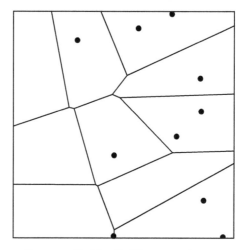

Figure 5.2: A power diagram with 10 cells. The y_j are the same as in figure 5.1, but prices are adjusted so that the areas of the cells are equal to one-tenth of the area of the square.

overall utility of inhabitants. Hence, one may look for

$$\max \mathbb{E}_P \left[\Phi \left(X, T(X) \right) \right]$$
$$\text{s.t. (5.12)}$$

(5.13)

and the problem is the following one.

Question 1 (Optimality). How should the optimal assignment T of fountains to individuals be determined?

If the authorities are unable or unwilling to use coercion, the alternative solution is classically the decentralized solution by price adjustment. In this case, a system of prices w_j is introduced, and prices adjust so that the market clearing constraint,

$$P\left(\mathcal{X}_j^v\right) = q_j \quad \forall j \in \{1, \dots, M\},$$

(5.14)

holds for a set of equilibrium prices v. In this case, the assignment of consumers to fountains is not decided by the central planner, but results from the individual rationality of consumers choosing the fountain that maximizes their utilities given transportation costs and market prices. In this case, the assignment is then given by $T(x) = \nabla u(x)$, where u is given by (5.5). It is to be expected that the market dynamics will be such that the prices of overdemanded fountains will rise, and the prices of underdemanded fountains will decrease. We thus ask this question:

Question 2 (Equilibrium). Is there a system of prices such that the market clears, that is, such that market clearing constraint (5.14) holds?

It will turn out that the two questions are intimately related; they are actually equivalent. It will not come as a surprise that the key to answering both questions is the Monge–Kantorovich theorem. In the sequel, we shall investigate these two questions in turn, and show that both of them can be addressed by means of Monge–Kantorovich theory.

5.2.1 Optimality Problem

We first consider question 1, which is the determination of an optimal assignment. The optimal coupling between inhabitants and fountains $(X, Y) \sim \pi$ is such that

$$\max_{\pi \in \mathcal{M}(P,Q)} \mathbb{E}_\pi \left[X'Y \right], \tag{5.15}$$

whose dual problem is

$$\min_{u,v} \mathbb{E}_P \left[u(X) \right] + \mathbb{E}_Q \left[v(Y) \right]$$
$$\text{s.t. } u(x) + v(y) \geq x'y, \tag{5.16}$$

where the constraint should hold almost surely with respect to P and Q. Hence, the constraint should be verified for $y \in \{y_1, \ldots, y_M\}$, and the constraint implies that $u(x)$ and $(v_j) := \left(v\left(y_j \right) \right)$ are related by expression (5.5). Thus, the dual Monge–Kantorovich problem (5.16) can be rewritten after substitution of $u(x)$ as

$$\min_{v \in \mathbb{R}^M} \mathbb{E}_P \left[\max_{j \in \{1,\ldots,M\}} \left\{ X'y_j - v_j \right\} \right] + \sum_{j=1}^M q_j v_j, \tag{5.17}$$

which is the finite-dimensional minimization problem of $S(v)$ over $v \in \mathbb{R}^M$, where S is given by

$$S(v) := W(v) + q'v. \tag{5.18}$$

Note that S is a convex function, thus the minimization problem considered is a convex minimization problem. Further, each individual x chooses a fountain at location $T(x) = \nabla u(x)$, which, by the first-order conditions in (5.17), satisfies the market clearing constraints (5.14). Hence question 1 above is fully answered. We summarize these results in the following theorem.

THEOREM 5.2. *The optimal coupling for problem (5.15) is given by* (X, Y), *where*

$$Y = \nabla u(X) \tag{5.19}$$

with

$$u(x) = \max_{j \in \{1,...,M\}} \{x' y_j - v_j\} \tag{5.20}$$

and the v_j are the solution to

$$\min_{v \in \mathbb{R}^M} \mathbb{E}_P \left[\max_{j \in \{1,...,M\}} \{X' y_j - v_j\} \right] + \sum_{j=1}^{M} q_j v_j. \tag{5.21}$$

With respect to the general Monge–Kantorovich theorem 2.2, theorem 5.2 yields the existence and the structure of the dual solution, which as the reader will recall, is not granted at the level of generality assumed in the former result. Here, the existence of a minimizer in the dual Monge–Kantorovich problem is granted because the latter boils down to the minimization of a convex function over a finite-dimensional vector space. Further, because of the shape of the surplus, the solution $u(x)$ of the dual is a convex, piecewise affine function. As convex, piecewise affine functions are almost everywhere differentiable, it follows that the set of fountains chosen by a given inhabitant is almost surely unique. Hence, the Monge problem (5.13) has a solution which coincides with the solution to the primal Monge–Kantorovich problem (5.15).

As we shall see in the next chapter, these results are typical of the scalar-product surplus (5.3). Anticipating the results in that chapter, we shall see that when the probability distributions of both X and Y are continuous, then the solutions $u(x)$ and $v(y)$ to the dual Monge–Kantorovich problem exist and are convex, although they are not piecewise affine in general. And we will see that the Monge problem also has a solution, which of course coincides with the solution to the Monge–Kantorovich problem, and which is still of the form $(X, Y = \nabla u(X))$.

5.2.2 Equilibrium Problem

We now consider the second question, namely, the question of the existence of an equilibrium. It should not come as a surprise by now that theorem 5.2 provides us with an answer to question 2 above, as well as an explicit determination of the equilibrium prices. Indeed, the partial derivative of $S(v)$ with respect to the price of fountain k is expressed as

$$\frac{\partial S(v)}{\partial v_k} = q_k + \frac{\partial W(v)}{\partial v_k} = q_k - P(\nabla u(X) = y_k);$$

thus the partial derivative $\partial S(v)/\partial v_k$ is interpreted as the excess supply of fountain k. Now take a vector v solution to (5.21), that is, a minimizer of S. The first-order conditions $\partial S(v)/\partial v_k = 0$ imply $q_k = P(\nabla u(X) = y_k)$ for all k, so that market clearing condition (5.14) holds with this vector v. Conversely, if a vector v satisfies the market clearing conditions, it also satisfies the first-order conditions, and by convexity of S, it is therefore a minimizer of S. Therefore we can summarize these findings in a formal statement as follows.

THEOREM 5.3. *Consider a price vector v optimal for problem (5.21). Then v is an equilibrium price vector, in the sense that $P(\mathcal{X}_k^v) = q_k$ for each k. Conversely, any equilibrium price vector is optimal for problem (5.21).*

5.3 COMPUTATION VIA CONVEX OPTIMIZATION

We turn to a discussion on the numerical determination of the prices (we discuss the determination of the v values, as the expression for the w values immediately follows). Since the function S to be minimized is convex, we can use a standard gradient descent algorithm.

ALGORITHM 5.4 (Standard gradient descent). *Take an initial guess of v^0. At step t, define v^{t+1} by*

$$v_j^{t+1} = v_j^t - \varepsilon_t \frac{\partial S}{\partial v_j}(v^t),$$

for ε_t small enough. Stop when $\partial S/\partial v_j(v^{t+1})$ is sufficiently close to 0.

Thus, the increase in prices is given by

$$v_j^{t+1} - v_j^t = \varepsilon \left(P\left(\nabla u(X) = y_k \right) - q_k \right), \tag{5.22}$$

which has an immediate economic interpretation: the fountains that are overdemanded *raise* their prices, while the fountains that are underdemanded *lower* their prices. This is a *tâtonnement process*.

Numerically, the tâtonnement algorithm simply requires the computation of $P(\nabla u(X) = y_k)$ for a set of prices (v_k). Equivalently, we need to evaluate the measure under distribution P of the cells $\mathcal{X}_k^v$. These cells are polyhedra, and they are described by (5.7) as intersections of half-spaces. There are standard computational geometry routines that compute the Euclidean volume of such polyhedra. Hence, when P is the Lebesgue measure, the problem can be addressed using existing tools. When P is not the Lebesgue measure, there are no off-the-shelf tools to compute the measure of the polyhedra. In that case, simulation methods should be used.

PROGRAMMING EXAMPLE 5.1. The package transport [136] provides useful resources for numerical optimal transportation, and in particular, power diagrams. The program powerdiagrams.R provides an example of equilibrium price determination, when P is uniform over $[0, 1]^2$. Initially the prices $\tilde{v}$ are set to 0, and the initial demand configuration is a Voronoi diagram.

The prices adjust over time so that the areas of the demand cells adjust to supply. Here, y1, y2, and vtilde are vectors of size 10 which represent, respectively, the first and second coordinates, and the prices of each of the 10 fountains. The instruction

```
pwd=power_diagram(y1,y2,vtilde,rect=c(0,1,0,1))
```

computes the power diagram, which is a list of cells, represented by their extreme points, while

```
plot(pwd,weights=FALSE)
```

plots the power diagram (if vtilde=0 this is the Voronoi diagram). For each cell j, set cellj=pwd$cells[[j]], and demand is computed using

```
demand[j]=polyarea(cellj[,1],cellj[,2]),
```

where polyarea, from the geometry package [75], computes the area of a cell. Finally, the prices are adjusted using

```
vtilde=vtilde-0.1*(demand-q)
```

where 0.1 is a tuning parameter which controls the speed of adjustment of the process, and the configuration converges to equilibrium.

5.4 EXERCISES

Exercise 5.1 (M). ***Power diagram as a Voronoi tessellation.*** *Show that any power diagram in $\mathbb{R}^d$ can be reexpressed as a Voronoi diagram in $\mathbb{R}^{d+1}$ projected onto a d-dimensional subspace.*

Exercise 5.2 (M). ***Logit as a power diagram.*** *Assume d = M, that is, the number of y values is equal to the dimension of the ambient space, and assume that y_j is the jth element of the canonical basis of $\mathbb{R}^M$, that is, all the components of y_j are 0 except for the jth entry, whose value is 1. Assume that P is the distribution of the centered Gumbel (extreme value type 1) distribution, that is,*

$$F_P(x) = \prod_{j=1}^{m} \exp\left(-\exp\left(-(x + \gamma)\right)\right),$$

where $\gamma \simeq 0.5772\ldots$ is Euler's constant. Show that W, defined by (5.10), is given by

$$W(v) = \log \sum_{k=1}^{M} \exp\left(-v_k\right),$$

and compute the equilibrium price vector v, as well as the value of the social planner's problem. Interpret in terms of discrete choice models.

Exercise 5.3 (E). ***Matching is a game.*** *This exercise requires some notions about supermodular games. Set $v_1 = 0$ and consider the (M − 1)-player game* where the profit of agent $k \in \{2, \ldots, M\}$ is given by

$$\Pi_k (v_k; v_{-k}) := -W(v) - q_k v_k,$$

where W is defined by (5.10).

(i) Show that $\Pi_k (v_k; v_{-k})$ is concave in v_k and supermodular.

(ii) Show that the best reply dynamics given by

$$v_k^{t+1} = \arg \min_{\bar{v} \in \mathbb{R}} W \left(\bar{v}; v_{-k}^t \right) + q_k \bar{v} \tag{5.23}$$

converges isotonically to the equilibrium price $(v_k^*)_k$ if the v_k^0 are chosen high enough. This suggests an algorithm for computing equilibrium prices which is an alternative to the one described in section 5.3. Discuss possible advantages of the present method.

Exercise 5.4 (C). *Optimal quantization.* Let P be the uniform distribution over $\mathcal{X} = [0, 1]^2$. For $\mathcal{Y} = \{y_1, \ldots, y_n\}$ a sample of n points in $\mathbb{R}^2$, let $Q^{y,q}$ be the distribution over the sample which assigns probability mass q_k on each y_k. Let $W(y, q)$ be the value of the Monge–Kantorovich problem

$$W (y, q) := \min_{\pi \in \mathcal{M}(P, Q^{y,q})} \mathbb{E}_\pi \left[\|X - Y\|^2 \right],$$

and consider the optimal quantization problem

$$\min_{\substack{y \in \mathbb{R}^{2n}, \\ q \geq 0 : \sum_{k=1}^{n} q_k = 1}} W (y, q).$$

(i) Show that if (y, q) is a solution and π is the associated optimal transportation plan, then $y_k = \mathbb{E}_\pi [X | Y = y_k]$.

(ii) Write a program to find the support points $y_1, \ldots, y_n$ in $\mathbb{R}^2$ along with the probability masses q_n so that y and q minimize $W (y, q)$.

Exercise 5.5 (E). *A Hotelling location game.* Consider a game where each fountain k is free to set its own price v_k. Assume that the profit of fountain k is determined by $\Pi_k(v_k, v_{-k}) = v_k \Pr(\mathcal{X}_k^v)$. Characterize the equilibrium conditions in this game. Fully solve for the symmetric equilibrium in the setting of exercise 5.2.

Exercise 5.6 (C). *Dynamics of the Hotelling game.* In the previous exercise, write down a program to study the best reply dynamics.

5.5 REFERENCES AND NOTES

Classical references on power diagrams are the various surveys by Aurenham-mer [7, 8] and references therein. A classical reference for the characteristics approach to demand is Lancaster [92]. In an industrial organization context, see the monograph by Anderson, de Palma, and Thisse [3], and the paper by Feenstra and Levinsohn [54]. See an application to political economy in Fryer and Holden [57]. The determination of prices to fit prescribed capacities by a convex optimization problem was initiated by Aurenhammer, Hoffmann, and Aronov [9]. For the link with the tâtonnement process, see Ekeland, Galichon, and Henry [51]. Exercise 5.3 appears to be original.

6

Quadratic Surplus

In this chapter, we assume as in the previous one that $\mathcal{X}$ and $\mathcal{Y}$ are convex subsets of $\mathbb{R}^d$, and that the surplus function is the scalar product, that is, $\Phi(x, y) = x'y$. However, we now assume only that P and Q are two general probability distributions, while we had previously considered a particular case of this situation when P has a density and Q has finite support. Hence, the setting we consider in this chapter is more general than in the previous one. Despite the increased level of generality, we will be able to provide virtually as strong results as before. Indeed, we shall show that, as before, the dual Monge–Kantorovich problem has a solution, as well as the Monge problem. While some of the structure of the previous chapter will be lost (in particular the link with piecewise affine functions and the link with power diagrams), we shall see that a remarkable new structure appears: convex duality. Indeed, it will turn out that if (u, v) is a solution to the dual Monge–Kantorovich problem under these assumptions, then u and v are "dual" in the sense of convex analysis, which we recall below. Further, the solution $T(x)$ to the Monge problem can be related to u in the sense that T is a solution to the Monge problem if and only if $T(x)$ is included in the subdifferential of u at x. As usual, these notions will be defined as they appear. Appendix D also has a brief reminder on convex analysis; see also Villani's short presentation in [148, section 2.1.3]. We believe that optimal transport provides the best possible introduction to convex analysis.

6.1 CONVEX ANALYSIS FROM THE POINT OF VIEW OF OPTIMAL TRANSPORT

We will begin with an informal presentation that intends to introduce the main concepts of convex analysis from the point of view of the theory of optimal transport and the underlying economic intuition. The Monge–Kantorovich theorem provides assumptions under which the value of the primal problem

$$\mathcal{W} = \sup_{\pi \in \mathcal{M}(P,Q)} \mathbb{E}_\pi \left[X'Y \right] \tag{6.1}$$

coincides with the value of the dual

$$W = \inf_{u(x)+v(y)\geq x'y} \mathbb{E}_P\left[u(X)\right] + \mathbb{E}_Q\left[v(Y)\right]. \tag{6.2}$$

Note, however, that assumptions are needed to make this statement valid. Indeed, theorem 2.2 required Φ to be bounded above, which is not the case for $\Phi(x, y) = x'y$ unless we assume that P and Q have bounded support. Another way to tackle the problem is to work instead with $\Phi(x, y) = -\frac{1}{2}|x - y|^2$, in which case we should assume that P and Q have finite second moment and replace $u(x)$ by $u(x) + \frac{1}{2}|x|^2$, and v by a similar quantity. We shall put aside these concerns for the moment; they will be addressed in the next section containing formal results.

First, we shall discuss the consequences of the existence of a solution (u, v) to the dual Monge–Kantorovich problem. Assume that a dual minimizer (u, v) exists; theorem 6.3 below provides a set of assumptions under which this is the case. If needed, redefine u and v so that they take the value $+\infty$ outside their supports, which are assumed to be convex. As argued in proposition 2.3, u and v are then related by

$$v(y) = \max_{x\in\mathbb{R}^d}\left\{x'y - u(x)\right\}, \tag{6.3}$$

$$u(x) = \max_{y\in\mathbb{R}^d}\left\{x'y - v(y)\right\}; \tag{6.4}$$

hence we see immediately that if (u, v) is a solution to the dual problem, then u and v are convex functions. Further, the expression of v as a function of u is the same as the expression of u as a function of v. This expression is a fundamental tool in convex analysis: it is called the *Legendre–Fenchel transform*, which is defined in general as follows.

DEFINITION 6.1 (Legendre–Fenchel transform). *The Legendre–Fenchel transform of f is defined by*

$$f^*(y) = \sup_{x\in\mathbb{R}^d}\left\{x'y - f(x)\right\}. \tag{6.5}$$

Some basic properties of Legendre–Fenchel transforms are given in proposition D.11 in the appendix. Using definition 6.1, expressions (6.3), (6.4) can be recast as $v = u^*$ and $u = v^*$. Hence u and v are *conjugate functions*, in the sense of convex analysis.

We now restate the demand sets of workers and firms in terms of subdifferentials of convex functions. For this, let us recall the basic economic interpretation of relations (6.3), (6.4), which we had previously spelled out: Expression (6.3) captures the problem of a firm of type y, which hires a worker x who offers the best trade-off between production if hired by y (i.e., $\Phi(x, y) = x'y$) and wage $u(x)$. Thus, firm y will be willing to match with any worker within the set of maximizers of (6.3), while worker x will be willing to match with any firm within the set of maximizers of (6.4). The set of maximizers of (6.3) and of (6.4) are called *subdifferentials* of v and u, respectively, where the subdifferential is formally defined as follows.

DEFINITION 6.2 (Subdifferential). *Let $f : \mathbb{R}^d \to \mathbb{R}$. The subdifferential of f at x, denoted $\partial f(x)$, is the set of $y \in \mathbb{R}^d$ such that for all $\tilde{x} \in \mathbb{R}^d$, $f(\tilde{x}) \geq f(x) + y'(\tilde{x} - x)$.*

Note that the definition does not require f to be convex; however, if f is convex, definition 6.2 immediately implies

$$\partial f(x) = \arg \max_{y} \left\{ x'y - f^*(y) \right\},\qquad (6.6)$$

which in turn implies that the subdifferential of a convex function is always nonempty (while the subdifferential of a nonconvex function can be empty in general). It also follows that if f is a convex function, the following statements are equivalent:

$$f(x) + f^*(y) = x'y,\qquad (6.7)$$

$$y \in \partial f(x),\qquad (6.8)$$

$$x \in \partial f^*(y).\qquad (6.9)$$

Going back to our worker–firm example, this has a straightforward economic interpretation. If worker x chooses firm y, then y maximizes $x'\tilde{y} - u^*(\tilde{y})$ over $\tilde{y}$, thus $y \in \partial u(x)$. This means that while worker x's equilibrium wage $u(x)$ is in general greater than or equal to the value $x'y - u^*(y)$ that the worker can extract from firm y, those two values necessarily coincide if x and y are willing to match, in which case $u(x) + u^*(y) = x'y$.

These considerations allow us to relate the solutions to the primal and dual problems. Recall that in the finite-dimensional case studied in chapter 3, the primal and dual problems are related by the complementary slackness condition (3.5). In the present case, let $(X, Y) \sim \pi$ be a solution to the primal problem, and (u, u^*) be a solution to the dual problem. Then almost surely X and Y are willing to match, which, by the previous discussion, implies

$$u(X) + u^*(Y) = X'Y,\qquad (6.10)$$

or equivalently $Y \in \partial u(X)$ or in turn $X \in \partial u^*(Y)$. In other words, the support of π is included in the set $\left\{ (x, y) : u(x) + u^*(y) = x'y \right\}$. This condition appears as the correct generalization of the complementary slackness condition in the finite-dimensional case. Unsurprisingly, taking the expectation with respect to π of equality (6.10) yields equality between the value of the dual problem on the left-hand side, and the value of the primal problem on the right-hand side.

More can be said when u is differentiable at x. In that case, it is not hard to show that $\partial u(x) = \{\nabla u(x)\}$, that is, it contains only one point, which is $\nabla u(x) = (\partial u(x)/\partial x_i)_i$, the vector of partial derivatives of u, or gradient of u. Similarly, if u^* is differentiable at y, then $\partial u^*(y) = \{\nabla u^*(y)\}$. Hence, if u and v are differentiable, then the equivalence between (6.8) and (6.9) implies $y = \nabla u(x)$ if and only if $x = \nabla u^*(x)$, that is,

$$(\nabla u)^{-1} = \nabla u^*. \tag{6.11}$$

Alternatively, relation (6.11) can be seen as a duality between first-order conditions and the envelope theorem. First-order conditions in the firm's problem (6.3) imply that if worker x is chosen by firm y, then $\nabla u(x) = y$, but the envelope theorem implies that the gradient in y of the firm's indirect profit $u^*(y)$ is given by $\nabla u^*(y) = x$, where x is chosen by y. Thus the first-order conditions and the envelope theorem are "conjugate" in the sense of convex analysis.

While we have spent some time discussing the case when the Kantorovich potentials u and v are differentiable, there is no a priori guarantee that they are so. In fact, chapter 5 provides an example of nondifferentiable Kantorovich potential, because $u(x)$ is piecewise affine. However, an important result in analysis, called Rademacher's theorem, implies that the set of nondifferentiable points of a convex function is of zero Lebesgue measure, and hence can be ignored for practical purposes as soon as P is continuous. This is typically the case in chapter 5, where we refer to $\nabla u(x)$ although it does not exist everywhere (and in particular, it does not exist when x lies on the boundary between several cells); however, the set where the gradient does not exist, which is the set of boundaries between cells, is of zero Lebesgue measure. Thus the Monge map solution, $T(x)$, can be defined as $T(x) = \nabla u(x)$ wherever the latter quantity exists, and $T(x)$ can be defined arbitrarily elsewhere, without affecting the distributional properties of $T(X)$.

6.2 MAIN RESULTS

We shall now give rigorous results. The first result deals with the existence of a solution to the dual problem, under the assumption of finite second moments for P and Q.

THEOREM 6.3. *If P and Q have finite second moments, then there exists a pair (u, v), a solution to the dual Monge–Kantorovich problem*

$$\inf_{u(x)+v(y)\geq x'y} \mathbb{E}_P\left[u(X)\right] + \mathbb{E}_Q\left[v(Y)\right].$$

Further, u and v are convex functions, and are mutually conjugate: $u = v^*$ and $v = u^*$.

We will not give a proof of this result, and will refer to [148, theorem 2.9] instead.

As argued in the previous paragraph, the existence of a pair of dual minimizers allows us to characterize the primal problem in much more detail. First, let us provide the formal statement which provides a generalization of the complementary slackness condition in finite dimensions. The intuition behind this result should be clear given the discussion in the previous paragraph.

PROPOSITION 6.4 (Knott–Smith). *Let $\pi \in \mathcal{M}(P, Q)$ and u be a convex function. Then π and (u, u^*) are respective solutions to the primal (6.1) and the dual (6.2) Monge–Kantorovich problems if and only if*

$$u(x) + u^*(y) = x'y \text{ holds for } \pi\text{-almost all } (x, y). \tag{6.12}$$

PROOF. Assume that (6.12) holds. Then, note that (u, u^*) satisfies the constraints of the dual; further, taking expectations with respect to π yields $\mathbb{E}_P[u(X)] + \mathbb{E}_Q[v(Y)] = \mathbb{E}_\pi[X'Y]$, which implies that π is an optimal primal solution and (u, u^*) is an optimal dual solution. Conversely, assume that π is an optimal primal solution and (u, u^*) is an optimal dual solution. Then $\mathbb{E}_\pi[u(X) + u^*(Y) - X'Y] = 0$, but $(x, y) \to u(x) + u^*(y) - x'y$ is nonnegative; thus (6.12) holds. $\square$

Recall that $u(x) + u^*(y) = x'y$ is equivalent to $y \in \partial u(x)$ or $x \in \partial u^*(y)$. Note that u is convex, and thus, by the theorem of Rademacher mentioned above, the set of points where u is not differentiable has zero Lebesgue measure. Hence, if P is continuous, u is P-almost surely differentiable, thus if $X \sim P$, one may identify $\partial u(X)$ with $\{\nabla u(X)\}$, and $Y \in \partial u(X)$ almost surely coincides with $Y = \nabla u(X)$. This remark implies that the Monge problem has a solution in this setting, which is the gradient of a convex function. We are thus able to state the following result.

THEOREM 6.5. *Assume that P and Q have finite second moments, and that P is continuous. Then, there is a unique (up to a constant) convex function u such that $X \sim P$ implies (i) $\nabla u(X) \sim Q$ and (ii) $(X, \nabla u(X))$ is the unique optimal coupling for problem (6.1).*

As a consequence, ∇u is the unique solution to the Monge problem

$$\sup_{T\#P=Q} \mathbb{E}_P[X'T(X)].$$

The proof, given in [148, theorem 2.12] is based on Rademacher's theorem, which implies that $\nabla u(x)$ exists Lebesgue-almost everywhere. Because P is continuous, $\nabla u(x)$ exists P-almost everywhere. Hence, if $(X, Y) \sim \pi$ is a solution to (6.1), then $Y = \nabla u(X)$ almost surely.

Of course, when both P and Q are continuous, this works both ways: the optimal coupling may be interpreted as $(X, \nabla u(X))$ and as $(\nabla u^*(Y), Y)$ alike. Hence, $X = \nabla u^*(Y)$ and $Y = \nabla u(X)$, and we are led to the following corollary.

COROLLARY 6.6. *Assume that P and Q have second moments and are continuous, and define u as in theorem 6.5 above. Then for P-almost every x and Q-almost every y, $\nabla u^* \circ \nabla u(x) = x$ and $\nabla u \circ \nabla u^*(y) = y$, and ∇u^* is the unique gradient of a convex function that pushes forward Q to P.*

Let us provide an example. Although there is no closed-form formula for the transportation potential between P and Q in dimensions greater than 1, there is one example of interest where there is actually a closed form, when both measures P and Q are Gaussian. Recall that the square root of a symmetric, positive-definite matrix S is the unique symmetric, positive-definite matrix $S^{1/2}$ such that $S = S^{1/2}S^{1/2}$.

Example 6.1. *Assume $P \sim \mathcal{N}(0, \Sigma_P)$ and $Q \sim \mathcal{N}(0, \Sigma_Q)$, where Σ_P and Σ_Q are two symmetric, positive-definite matrices. Then the solution (u, v) to the dual Monge–Kantorovich problem is given by $u(x) = \frac{1}{2}x'S_u x$ and $v(y) = \frac{1}{2}y'S_v y$, where*

$$S_u = \Sigma_P^{-1/2}\left(\Sigma_P^{1/2}\Sigma_Q\Sigma_P^{1/2}\right)^{1/2}\Sigma_P^{-1/2}$$

and

$$S_v = \Sigma_Q^{-1/2}\left(\Sigma_Q^{1/2}\Sigma_P\Sigma_Q^{1/2}\right)^{1/2}\Sigma_Q^{-1/2}$$

(note that $S_v = S_u^{-1}$), and the optimal assignment is $T(x) = S_u x$; the value of the Monge–Kantorovich problem is

$$\mathcal{W} = \mathrm{Tr}\left[\left(\Sigma_P^{1/2}\Sigma_Q\Sigma_P^{1/2}\right)^{1/2}\right].$$

Indeed, $u(x)$ is convex because S is positive definite, symmetric and $T(x) = \nabla u(x) = Sx$ sends the probability measure $\mathcal{N}(0, \Sigma_P)$ to the probability measure $\mathcal{N}(0, \Sigma_Q)$; the expression for $\mathcal{W}$ follows. In particular, in the case $P = \mathcal{N}(0, I_d)$, then $\mathcal{W} = \mathrm{Tr}[\Sigma_Q^{1/2}]$ which in matrix analysis is called the trace norm *of Σ_Q. When $d = 2$, letting $\sigma_1^2 = \mathrm{Var}(Y_1)$, $\sigma_2^2 = \mathrm{Var}(Y_2)$, and $\varrho = \mathrm{corr}(Y_1, Y_2)$, we have the formula $\mathrm{Tr}[\sqrt{\Sigma}]^2 = \mathrm{Tr}(\Sigma) + 2\sqrt{\det \Sigma}$, so we get the explicit expression*

$$\mathcal{W} = \sqrt{\sigma_1^2 + \sigma_2^2 + 2\sigma_1\sigma_2\sqrt{1 - \varrho^2}}.$$

6.3 VECTOR QUANTILES

The previous results rely on a somewhat strong assumption, namely, that P and Q have finite second moments. Assuming finite second moments ensures that the objective function $\mathbb{E}_\pi\left[-\frac{1}{2}(X-Y)^2\right]$ remains finite for any $\pi \in \mathcal{M}(P,Q)$. However, the condition is not needed for a large part of the results to hold, as understood by McCann, who proved the following fundamental result.

THEOREM 6.7 (McCann). *Let P and Q be two probability measures over $\mathbb{R}^d$, such that P is continuous. Then there is a measurable map T which is the gradient of some convex function u, and such that T#P = Q. Further, T is unique in the sense that two such maps coincide P-almost everywhere.*

PROOF. See the proof of [148, theorem 2.32]. □

In other words, if $X \sim P$, there is a unique (up to an additive constant) convex function u such that $\nabla u(X) \sim Q$. Hence, this theorem provides a canonical representation of the distribution Q of a random vector as a particular transformation of a given reference measure P. Note that this theorem does not refer to the Monge–Kantorovich problem; as P and Q may not have second moments, this problem is not well defined.

Theorem 6.7 provides a far-reaching generalization of the notion of a quantile transform seen in expression (4.2). Indeed, if $d = 1$ and $P = \mathcal{U}([0,1])$, then the gradient of a convex function u is simply the derivative of a convex function, which is a nondecreasing function. Hence $\nabla u(x) = u'(x)$ is the nondecreasing function that maps the uniform distribution on $[0,1]$ onto Q. Therefore, $u'(x) = F_Q^{-1}(x)$. More generally, if P is some continuous distribution on the real line, then $u'(x) = F_Q^{-1} \circ F_P(x)$. The above analogy with a univariate quantile leads to the following definition.

DEFINITION 6.8 (Vector quantile). *Let P be a continuous probability measure on $\mathbb{R}^d$. Given a probability measure Q on $\mathbb{R}^d$, the P-vector quantile associated with Q is the unique gradient of a convex function T such that T#P = Q.*

Of course, when $d = 1$ and P is the uniform distribution over $[0,1]$, this notion coincides with the classical notion of a quantile. One of the advantages of this notion of vector quantiles is that it allows us to define *empirical quantiles*. Assume that an i.i.d. sample of Q, denoted $\{y_1, \ldots, y_n\}$, is observed, and let Q_n be the corresponding empirical distribution. There is a convex function u_n such that $\nabla u_n(X) \sim Q_n$ as soon as $X \sim P$; ∇u_n is the optimal map from continuous distribution P to finite distribution Q_n associated with the scalar-product surplus. Hence, by the results of chapter 5, one is able to state the following definition.

DEFINITION 6.9 (Empirical vector quantiles). *Let P be a continuous probability measure on $\mathbb{R}^d$, and consider a sample of n vectors $\{y_1, \ldots, y_n\}$ in $\mathbb{R}^d$. The P-empirical vector quantile T_n associated with the sample is the map T_n such that $T_n(x) = \arg\max_k \{x'y_k - v_k\}$, where (v) is a solution of*

$$\min_v \mathbb{E}_P \left[\max_k \{X'y_k - v_k\} \right] + \frac{1}{n} \sum_{k=1}^n v_k.$$

A natural property to expect is consistency: when the sample size is large, the empirical vector quantile should converge to the true vector quantile. This is established in the following result.

THEOREM 6.10 (Consistency of empirical vector quantiles). *Assume that P and Q are continuous and compactly supported in $\mathbb{R}^d$. Let $\mathcal{X}_0$ and $\mathcal{Y}_0$ be the interiors of their respective supports. Let $T = \nabla u$ be the P-vector quantile of Q. Assume that u and u^* are everywhere differentiable on $\mathcal{X}_0$ and $\mathcal{Y}_0$ respectively, so that ∇u induces a homeomorphism of $\mathcal{X}_0$ onto $\mathcal{Y}_0$. Let T_n be the P-empirical vector quantile associated with an empirical sample of n i.i.d. draws from Q. Then*

$$T_n \to T$$

uniformly on compacts included in $\mathcal{X}_0$.

The proof of this result is omitted; the reader is referred to [37, theorem 3.1].

After consistency, it is natural to look for a central limit theorem. For this, we need to consider the empirical vector quantile process, $g_n(x) = n^{1/2} (T_n(x) - T(x))$, which extends the empirical quantile process in the scalar case. The asymptotic behavior of $g_n(x)$ is left as an open problem.

OPEN PROBLEM. *What is the asymptotic behavior of the empirical vector quantile process?*

While vector quantiles provide a natural generalization of the univariate notion of quantiles (see an application in section 9.4), a caveat is in order: they are not stable by composition in dimensions higher than 1, as argued in the following proposition.

PROPOSITION 6.11. *Consider three probability measures on $\mathbb{R}^d$, P, Q, and R, and assume that P and Q are continuous. Let T_1 be the P-vector quantile of Q, and T_2 be the Q-vector quantile of R, which means that $T_1 \# P = Q$ and $T_2 \# Q = R$, and both T_1 and T_2 are gradients of convex functions. Then,*

    (i) *if $d = 1$, then $T_2 \circ T_1$ is the P-vector quantile of R;*

    (ii) *however, if $d > 1$, then $T_2 \circ T_1$ is not the P-vector quantile of R in general.*

PROOF. If $d = 1$, then $T_1 = F_Q^{-1} \circ F_P$ and $T_2 = F_R^{-1} \circ F_Q$; thus $T_2 \circ T_1 = F_R^{-1} \circ F_P$ is the P-vector quantile of R.

If $d > 1$, consider $P = \mathcal{N}(0, I_d)$, $Q = \mathcal{N}(0, \Sigma_1)$, and $R = \mathcal{N}(0, \Sigma_2)$. Then, by example 6.1, $T_1 = \Sigma_1^{1/2}$ and $T_2 = \Sigma_1^{-1/2}(\Sigma_1^{1/2}\Sigma_2\Sigma_1^{1/2})^{1/2}\Sigma_1^{-1/2}$. However, $T_2 \circ T_1(x) = \Sigma_1^{-1/2}(\Sigma_1^{1/2}\Sigma_2\Sigma_1^{1/2})^{1/2}x$, while the P-vector quantile of R is $x \to \Sigma_2^{1/2}x$. These quantities are not equal in general. $\square$

6.4 POLAR FACTORIZATION

The results in the previous paragraph have a beautiful geometric interpretation, in connection with stochastic orders and multivariate analysis. Because the economic applications of this paragraph are only indirect, it may safely be skipped by readers who prefer to focus on more concrete applications of the theory.

THEOREM 6.12 (Brenier). *Let $\mathcal{X}$ and $\mathcal{Y}$ be two measurable subsets of $\mathbb{R}^d$. Let P be a probability measure on $\mathcal{X}$ which is continuous and has finite second moments. Let $\varphi : \mathcal{X} \to \mathcal{Y}$ be a map such that $\varphi\#P$ is continuous and has finite second moments. Then there is a unique pair of maps ∇u and τ such that*

> *(i) $\nabla u : \mathcal{X} \to \mathbb{R}^d$ is the gradient of a convex function;*
> *(ii) $\tau : \mathcal{X} \to \mathcal{X}$ preserves measure P; and*
> *(iii) φ is the composition of ∇u and τ, that is,*

$$\varphi = \nabla u \circ \tau. \tag{6.13}$$

The statement "$\varphi\#P$ is continuous" means that for any measurable subset B of $\mathcal{Y}$, $\mathcal{L}eb(B) = 0$ implies $P\left(\varphi^{-1}(B)\right) = 0$. Also, "$\tau$ preserves measure P" means that for any measurable subset B of $\mathcal{X}$, $P(\tau(B)) = P(B)$; for instance, the map $\tau(x) = 1 - x$ preserves the uniform measure on $[0, 1]$. We provide a proof of this result (see also [148, theorems 3.8, 3.15]). It consists of letting $Q = \varphi\#P$, and noting that, by theorem 6.5, there is a convex function u such that $P = \nabla u^*\#Q$. Thus $P = \nabla u^*\#(\varphi\#P) = (\nabla u^* \circ \varphi)\#P$. Therefore, $\tau = \nabla u^* \circ \varphi$ preserves measure P, and φ factorizes as (6.13).

PROOF OF THEOREM 6.12. Consider $X \sim P$, and call Q the distribution of $\varphi(X)$. Then $Q = \varphi\#P$, thus by assumption, Q is continuous with finite second moments. By theorem 6.5, there exists a convex function u such that for P-almost every x and Q-almost every y, $\nabla u^* \circ \nabla u(x) = x$ and $\nabla u \circ \nabla u^*(y) = y$, and ∇u^* is the unique gradient of a convex function that pushes forward Q to P. Hence

$$\nabla u^* \left(\varphi(X)\right) \sim P;$$

thus $\tau := \nabla u^* \circ \varphi$ preserves measure P, and $\varphi = \nabla u \circ \tau$. □

In particular, when $d = 1$ and $P = \mathcal{U}([0,1])$, this result boils down to a classical result called Ryff's theorem: any function φ such that $\varphi \# P$ is continuous and has finite second moments can be written as

$$\varphi = F_{\varphi \# P}^{-1} \circ \tau,$$

where $F_{\varphi \# P}$ is the c.d.f. of $\varphi \# P$, and $\tau = F_{\varphi \# P} \circ \varphi$ preserves Lebesgue measure on $[0, 1]$; $F_{\varphi \# P}^{-1}$ is called the *increasing rearrangement* of φ.

A way to interpret theorem 6.12 using a variational problem consists of letting $X \sim P$ and arguing that τ in (6.13) is the solution to the L^2-minimization problem

$$\min_{\tilde{\tau}:\tilde{\tau}\#P=P} \mathbb{E}\left[|\varphi(X) - \tilde{\tau}(X)|^2 \right], \tag{6.14}$$

or equivalently,

$$\max_{\tilde{\tau}:\tilde{\tau}\#P=P} \mathbb{E}\left[\varphi(X)'\tilde{\tau}(X) \right],$$

which is a Monge problem between the distribution P and itself, associated with surplus $\Phi(x, y) = \varphi(x)'y$. It is easy to see that the solution to this problem is provided by $\tau = \nabla u^* \circ \varphi$, where u has been characterized above. Indeed, letting $Y = \varphi(X)$ and calling Q its distribution, and $Y^{\tilde{\tau}} = \nabla u(\tilde{\tau}(X))$, one has $Y^{\tilde{\tau}} \sim Q$, and

$$\mathbb{E}\left[Y'\tau(X)\right] = \mathbb{E}\left[\varphi(X)'\nabla u^*(\varphi(X))\right] = \mathbb{E}\left[Y'\nabla u^*(Y)\right]$$
$$\geq \mathbb{E}\left[Y'\nabla u^*\left(Y^{\tilde{\tau}}\right)\right] = \mathbb{E}\left[Y'\nabla u^*(\nabla u(\tilde{\tau}(X)))\right] = \mathbb{E}\left[Y'\tilde{\tau}(X)\right],$$

where the central inequality again follows from theorem 6.5. Define the equidistribution class of P as follows.

DEFINITION 6.13. *The equidistribution class of P is the set of random vectors $\tilde{X}$ such that $\tilde{X} \sim P$.*

Therefore, we have the following straightforward, but important interpretation of τ in (6.13).

PROPOSITION 6.14. *Under the assumptions and notation of theorem 6.12, let $X \sim P$. Then $\tau(X)$ is the L^2-projection of $\varphi(X)$ on the equidistribution class of P.*

Note that the equidistribution class of P is not a convex set. For instance, if $X_1, \ldots, X_n$ are n i.i.d. random vectors with distribution P, then $n^{-1}(X_1 + \cdots + X_n)$ is in general not distributed as P. In fact, the convex closure of the equidistribution class of P is the set of random vectors with distributions that dominate P in the convex ordering (see, for example, Strassen [141, theorem 2] and references therein).

Polar factorization can be seen as a far-reaching nonlinear generalization of the polar factorization of real matrices. Recall that an invertible matrix A can be uniquely factorized into

$$A = SU, \tag{6.15}$$

where S is symmetric, positive definite, and U is orthogonal. Here, S and A are given by $S = (AA')^{1/2}$, and $U = (AA')^{-1/2} A$. This decomposition follows from theorem 6.12 by taking the standard multivariate normal distribution for P and the linear map associated with the matrix A for φ, that is, $P = \mathcal{N}(0, I_d)$ and $\varphi(x) = Ax$. Then, $\varphi = \nabla u \circ \tau$, where ∇u is the gradient of the convex function that maps P to $\varphi \# P$, and τ preserves P. However, it is not hard to see that $u(x) = \frac{1}{2} x' (AA')^{1/2} x$, which implies $S = (AA')^{1/2}$ and $\tau(x) = \nabla u^* (\varphi(x)) = (AA')^{-1/2} Ax$.

As seen in the previous discussion, theorem 6.12 provides a natural nonlinear generalization of a basic matrix decomposition result, which is the polar factorization of matrices (6.15). There is, however, another basic matrix decomposition result, which is the *bipolar factorization* of matrices, or singular value decomposition. Recall that a square real-valued matrix A can be decomposed as

$$A = U_1' D U_2, \tag{6.16}$$

where D is a diagonal matrix whose diagonal coefficients are nonnegative and in nondecreasing order, and U_1 and U_2 are orthogonal matrices. Note that this representation can be obtained from the eigenvalue decomposition of S in (6.15), which yields $S = U_1' D U_1$, where D is diagonal with nonnegative and nondecreasing diagonal coefficients, and U_1 is orthogonal. Setting $U_2 = U_1 U$ then yields (6.16).

It is natural to wonder what the nonlinear generalization of the singular value decomposition of matrices (6.16) is. However, this is still an open problem, which we state as such despite its somewhat vague formulation.

OPEN PROBLEM. *Polar factorization is the nonlinear generalization of the polar decomposition of matrices (6.15). What is the nonlinear "bipolar decomposition," that would provide a generalization of the singular value decomposition of matrices (6.16) in the same spirit?*

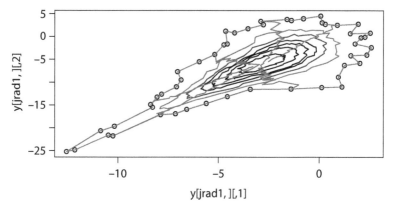

Figure 6.1: Outcome of programming example 6.1(a).

6.5 COMPUTATION BY DISCRETIZATION

Numerical computation of optimal transportation problems in $\mathbb{R}^d$ when P and Q are continuous is a topic under active investigation. We will not review the forefront of the research on this topic, which is too recent at this stage to allow for a big-picture perspective. Instead, we will describe the simplest of the methods, which consists of sampling P and Q (either randomly or in some regular manner) into discrete measures P_n and Q_n with n equally weighted atom points $\{x_1, \ldots, x_n\}$ and $\{y_1, \ldots, y_n\}$, and solving the discrete assignment problem between the x_i values and the y_i values, using one of the discrete assignment methods discussed in section 3.4. This method is implemented in programming example 6.1.

It is clear from figure 6.1 that the discretization method employed is quite crude. There are several issues with the approach. First, we have chosen to sample Q randomly; a more regular spacing could have been used instead. Second, the number of discretization points (2000) is clearly insufficient; unfortunately, a simple laptop cannot employ more discretization points and finish in a reasonable amount of time, but a more powerful computer could work with far more. Third, the choice of the discretization grid for P was arbitrary, with uniform weights; one may have gained in accuracy from adapting the grid to the target measure Q; see, for example, [106]. Fourth, alternative methods may be employed, such as entropic regularization, discussed in section 7.3; see, for example, [14].

PROGRAMMING EXAMPLE 6.1. In the program `VectorQuantile.R`, we follow Chernozhukov et al. [37], who compute by simulation an optimal transport map from P, the spherical distribution on the unit ball on $\mathbb{R}^2$ with uniform radius, to Q which is a Clayton distribution, whose c.d.f. is

$$F_Q(y) = \max \left(y_1^{-\theta} + y_2^{-\theta} - 1, 0 \right)^{-1/\theta}.$$

Let ∇u be the optimal transport from P to Q. Our goal is to draw

- the contour sets $\{\nabla u(x) : |x| = r\}$ for a number of values of $r \in [0, 1]$;
- the paths $\{\nabla u(tn) : t \in [0, 1]\}$ for a number of directions n such that $|n| = 1$.

In order to do this, the spherical distribution is sampled into a regular grid of 40 radius points between 0 and 1, and 50 angle points between 0 and 2π. A random sample of 2000 points is taken for Q. The optimal transport map ∇u is approximated by computing the optimal assignments of the discretized version of P to the random sample of Q. Optimal assignment is computed using the function `assignment` of the package `adagio` [21], which is fast but does not return the dual prices, which may be an inconvenience in some settings. The outcome is illustrated in figure 6.1.

6.6 EXERCISES

Exercise 6.1 (M). **Spherical invariance.** *Let P be the uniform distribution over the unit ball $B(1)$ of $\mathbb{R}^d$. Let Q be a distribution with spherical invariance, namely, if $Y \sim Q$ and R is an orthogonal matrix, then $RY \sim Q$. Let $\Phi(x, y) = x'y$. Let u and v be a solution pair to the dual problem, and $(X, Y) \sim \pi$, where π is an optimal coupling solution to the primal problem.*

- (i) *Show that there is $\zeta : \mathbb{R}^+ \to \mathbb{R}$ such that $u(x) = \zeta(|x|)$.*
- (ii) *Show that the map $x \to \zeta(|x|)$ is convex if and only if ζ is nondecreasing and convex.*
- (iii) *Show that $|Y| = \zeta'(|X|)$, and deduce an expression for ζ as a function of $F_{|Y|}$ and $F_{|X|}$, the c.d.f.s of $|Y|$ and $|X|$, respectively.*

Exercise 6.2 (M). **The Knothe–Rosenblatt map.** *Let P and Q be two continuous distributions over $\mathbb{R}^2$ with finite second moments. For $\lambda > 0$, let $T^\lambda(x)$ be the optimal transport map between P and Q relative to surplus $\Phi^\lambda(x, y) = x_1 y_1 + \lambda x_2 y_2$. Let $\bar{T}(x) = \lim_{\lambda \to 0^+} T^\lambda(x)$. Show that*

$$\bar{T}(x_1, x_2) = \left(\bar{T}_1(x_1), \bar{T}_2(x_1, x_2) \right),$$

where $\bar{T}$ is called the Knothe–Rosenblatt map *between distributions P and Q, and is defined by*

$$\bar{T}_1(x_1, x_2) = F_{Y_1}^{-1}\left(F_{X_1}(x_1)\right) \quad and$$

$$\bar{T}_2(x_1, x_2) = F_{Y_2|Y_1}^{-1}\left(F_{X_2|X_1}(x_2|X_1 = x_1)\,|\,Y_1 = F_{Y_1}^{-1}\left(F_{X_1}(x_1)\right)\right)$$

(note that $\bar{T}_1(x_1, x_2)$ does not depend on x_2).

Exercise 6.3 (C). *A numerical experiment. Based on the theoretical results of exercise 6.2 and on programming example 5.1, write a code that demonstrates the evolutions of the cells $T^\lambda(\{y\})$ when λ varies between 0 and $+\infty$, in the case when P is the uniform distribution over $[0, 1]^2$, and Q is an empirical distribution.*

Exercise 6.4 (C). *Smoothing the contours. Repeat the simulation in example 6.1 1000 times in order to provide a smoothed estimate of the contours.*

Exercise 6.5 (E). *Quadratic–Gaussian matching. Consider a population of workers whose skills are represented by a d-dimensional vector with distribution $P = \mathcal{N}(0, \Sigma_P)$. There is a population of firms of equivalent size that are also characterized by a d-dimensional vector of characteristics, distributed according to $Q = \mathcal{N}(0, \Sigma_Q)$. The surplus of matching employee x to firm y is $\Phi(x, y) = x'Ay$, where A is a $d \times d$ invertible matrix. Using the results in example 6.1, characterize the equilibrium assignment map and wage curve.*

Exercise 6.6 (E). *Hedonism. Consider a hedonic model, where a consumer with unobserved characteristics $\varepsilon \in \mathbb{R}^d$ chooses a good of quality $y \in \mathbb{R}^d$ whose price is $v(y)$ in order to maximize utility $\varepsilon' y - v(y)$ over y. The distribution P of ε is assumed to be known, and v is assumed to be convex. The distribution of choices y, denoted Q, is observed. Show that v is identified as a solution to the dual Monge–Kantorovich problem.*

6.7 REFERENCES AND NOTES

The statements of most of the results in this section are based on Villani's exposition in [148, chapter 2]. General references for section 6.1 on convex analysis are Rockafellar [122] and Ekeland and Temam [53]. Theorem 6.3, proposition 6.4, and theorem 6.5 were first discovered by Knott and Smith [87], and improved by Rüschendorf and Rachev [129]. The idea was independently rediscovered by Brenier in a series of papers [24, 25], and he was the first to make several important links with problems in mathematical physics, leading to a considerable renewal of interest for the field. This is why many authors (including Villani) associate Brenier's name with the result.

Example 6.1 is due to Olkin and Pukelsheim [115] and Dowson and Landau [46]. Theorem 6.7 is due to McCann [102]. References for vector quantiles are [51, 65, 37, 33]. In particular, theorem 6.10 is taken from [37]. A reference for the univariate empirical quantile process is [146]. Theorem 6.12 on polar factorization and its interpretation in proposition 6.14 are due to Brenier [25]; see a fascinating discussion in [148, chapter 3] . The reference for Ryff's theorem is [131]. Various numerical methods improving upon naive sampling techniques are investigated in Damien Bosc's PhD thesis [22]. A reference for programming example 6.1 is Chernozhukov et al. [37]. A reference for exercise 6.1 is [42]. Exercise 6.2 is based on [34]. Exercise 6.6 is based on [38].

—7—

More General Surplus

We now want to investigate existence and qualitative properties of solutions of the optimal transport problem still in the multivariate case, but with surpluses that are more general than the quadratic surplus that we have considered thus far. It turns out that, under assumptions which generalize the univariate submodularity assumption of chapter 4, as well as the scalar-product surplus of chapter 6, we are still able to say much, and the properties of the solutions resemble those derived in chapter 6 for quadratic surpluses— we just need to replace the standard notion of convexity by *generalized convexity*, or Φ-convexity.

7.1 GENERALIZED CONVEXITY

As in the previous chapter, we consider the case when $\mathcal{X} = \mathcal{Y}$ are closed subsets of $\mathbb{R}^d$, and we begin by studying the implications of the existence of a solution to the dual Monge–Kantorovich problem. Recall the result of proposition 2.3: if (u, v) is a solution to the dual Monge–Kantorovich problem, then

$$v(y) = \sup_{x \in \mathcal{X}} \{\Phi(x, y) - u(x)\} \quad \text{and} \quad u(x) = \sup_{y \in \mathcal{Y}} \{\Phi(x, y) - v(y)\}. \qquad (7.1)$$

In the case seen in the previous chapter, that is, when $\Phi(x, y) = x'y$, this implies that u and v form a pair of conjugate functions in the sense of convex analysis. However, much of the theory remains the same if one defines a generalized notion of conjugacy, under which u and v in (7.1) form a conjugate pair.

DEFINITION 7.1 (Φ-transform). *Consider a function $f : \mathcal{X} \to \mathbb{R}$. The Φ-convex conjugate of f, or Φ-transform of f, is defined as the function $f^{\Phi} : \mathcal{Y} \to \mathbb{R}$ such that*

$$f^{\Phi}(y) = \sup_{x \in \mathcal{X}} \{\Phi(x, y) - f(x)\}.$$

Note that it follows from expression (7.1) that $v = u^\Phi$. Letting

$$\Phi^\mathsf{T}(y, x) := \Phi(x, y),$$

we can conversely express u as the Φ^T-transform of v, that is, $u = v^{\Phi^\mathsf{T}}$. This motivates the following definition of Φ-convexity.

DEFINITION 7.2 (Φ-convexity). *A function $f : \mathcal{X} \to \mathbb{R}$ is called Φ-convex if it is not identically $+\infty$ and if there is a function $g : \mathcal{Y} \to \mathbb{R}$ such that $f = g^{\Phi^\mathsf{T}}$.*

The following properties of Φ-conjugacy are left as an exercise to the reader.

PROPOSITION 7.3. *The following properties hold true:*

(i) *If $f \le \tilde{f}$, then $f^\Phi \ge \tilde{f}^\Phi$.*
(ii) *If $\tilde{f} = f + c$, where $c \in \mathbb{R}$, then $\tilde{f}^\Phi = f^\Phi - c$.*
(iii) *$f(x) + f^\Phi(y) \ge \Phi(x, y)$.*
(iv) *$f^{\Phi\Phi^\mathsf{T}}(x) \le f(x)$, with equality if and only if f is Φ-convex.*
(v) *If f_1 and f_2 are Φ^T-convex, then $\max\{f_1, f_2\}$ is Φ-convex, and*

$$\max\{f_1, f_2\} = \min\{f_1^\Phi, f_2^\Phi\}^{\Phi^\mathsf{T}}.$$

Note that $f^{\Phi\Phi^\mathsf{T}}$ is the Φ-convex envelope of f, which has expression

$$f^{\Phi\Phi^\mathsf{T}}(x) = \sup_{y' \in \mathcal{Y}} \inf_{x' \in \mathcal{X}} \left\{ f\left(x'\right) + \Phi\left(x, y'\right) - \Phi\left(x', y'\right) \right\}.$$

The notion of a subdifferential function extends just as well. Recall that in standard convex analysis, $y \in \partial f(x)$ if and only if $f(x) + f^*(y) = x'y$. This motivates the following definition.

DEFINITION 7.4 (Φ-subgradient). *The Φ-subgradient of a function $f : \mathcal{X} \to \mathbb{R}$ is defined as*

$$\partial^\Phi f(x) = \{y \in \mathcal{Y} : f(x) + f^\Phi(y) = \Phi(x, y)\}.$$

Whenever there is no ambiguity, we shall use the abbreviated notation $\partial f(x)$.

Note that when f is Φ-convex,

$$\partial^\Phi f(x) = \arg\max_{y \in \mathcal{Y}} \left\{ \Phi(x, y) - f^\Phi(y) \right\}.$$

PROPOSITION 7.5. *Let $\Phi : \mathcal{X} \times \mathcal{Y} \to \mathbb{R}$ and $f : \mathcal{X} \to \mathbb{R}$.*

(i) *When Φ and f are differentiable in x,*

$$y \in \partial^\Phi f(x) \quad \text{implies} \quad \nabla f(x) = \nabla_x \Phi(x, y). \tag{7.2}$$

(ii) When Φ and f are twice differentiable in x,

$$y \in \partial^\Phi f(x) \quad \text{implies} \quad D^2 f(x) \succeq_{\text{spd}} D^2_{xx} \Phi(x, y). \tag{7.3}$$

Recall that $A \succeq_{\text{spd}} B$ means $A - B$ is symmetric, positive semidefinite.

PROOF. If $y \in \partial^\Phi f(x)$, then $x \in \arg\max_{x' \in \mathcal{X}} \Phi(x', y) - f(x')$; thus, (i) follows by first-order conditions, and (ii) follows by second-order conditions. $\qquad\square$

Both points made in this result are worth commenting on. Point (i) explains the importance of the condition that ensures the invertibility of $y \to \nabla_x \Phi(x, y)$ which will be given in theorem 7.7. Indeed, assume that this function is invertible, and let $p \to e_x(p)$ be this inverse; that is,

$$e_x(\cdot) := \nabla_x \Phi(x, \cdot)^{-1}, \tag{7.4}$$

so that $\nabla_x \Phi(x, e_x(p)) = p$. Then by (7.2), $y \in \partial^\Phi f(x)$ implies $y = e_x(\nabla f(x))$, so that whenever $\partial^\Phi f(x)$ is single valued, one has necessarily

$$\partial^\Phi f(x) = \{e_x(\nabla f(x))\}.$$

Point (ii) provides a differential view on Φ-convexity. In particular, it shows that Φ-convex functions are functions whose Hessian is greater than that of $\Phi(\cdot, y)$ in the semidefinite ordering. Note that in the case of standard convexity, $\Phi(x, y) = x'y$, thus the Hessian of $\Phi(\cdot, y)$ is 0, which recovers the standard differential characterization of convexity. Therefore, if Φ is convex in x, it is to be expected that Φ-convex functions are, in particular, convex functions in the usual sense; instead, if Φ is concave in x, it is to be expected that there will be functions that are Φ-convex functions yet that are not convex in the usual sense.

Let us now provide several examples. Our first example will obviously consist of the scalar-product surplus seen in the previous chapter.

Example 7.1 (Scalar-product surplus). *In the case $\Phi(x, y) = x'y$, all the notions coincide with their classical version in standard convex analysis. In this case $\Phi^\mathsf{T} = \Phi$.*

Example 7.2 (Convex displacement cost). *In the case $\Phi(x, y) = -k(x - y)$, where $k : \mathbb{R}^d \to \mathbb{R}$ is a cost function that is strictly convex and C^1, it is easily seen that*

$$\partial^\Phi f(x) = \{x - \nabla k^*(-\nabla f(x))\},$$

where k^* is the Legendre transform of k. In particular, when $k(z) = \frac{1}{2}|z|^2$, this becomes $\partial^\Phi f(x) = \{x + \nabla f(x)\}$, and f is Φ-convex as soon as $x \to f(x) + \frac{1}{2}|x|^2$ is convex.

Example 7.3 (Distance cost). *In the case* $\mathcal{X} = \mathcal{Y} \subseteq \mathbb{R}^d$ *and* $\Phi(x, y) = -d(x, y)$, *where d is a distance, then*

> (i) *the* Φ-*convex functions are the 1-Lipschitz functions;*
> (ii) *if f is* Φ-*convex, then* $f^{\Phi}(y) = -f(y)$;
> (iii) $\partial^{\Phi} f(x) = \{y : f(y) = f(x) + d(x, y)\}$.

This case is interesting for two reasons. The first one is historical: this version of the problem was initially considered by Monge in 1781. Second, the Monge–Kantorovich theorem particularizes in this setting to the Kantorovich–Rubinstein theorem, which states that

$$\inf_{\pi \in \mathcal{M}(P,Q)} \mathbb{E}_{\pi}\left[d(X, Y)\right] = \sup_{u \text{ is 1-Lipschitz}} \mathbb{E}_P\left[u(X)\right] - \mathbb{E}_P\left[u(X)\right].$$

We move on to other examples of some economic significance.

Example 7.4 (Intertemporal utility under uncertainty). *Assume that the states of the world are* $i = 1, \ldots, d$, *and that* $\mathcal{Y}$ *is the set of probabilities on the state of the world, that is,* $\mathcal{Y}$ *is the set of* $y \in \mathbb{R}^d$ *such that* $y \geq 0$ *and* $\sum_{i=1}^d y_i = 1$. *Let x be a financial portfolio that pays off* x_i *in state i, and let* $C(x)$ *be the cost (market price) of this portfolio. The set of portfolios is denoted* $\mathcal{X}$. *An investor has wealth W at time 0, expected utility preferences with utility function U, and time discount parameter* δ, *giving an intertemporal utility of*

$$\delta \sum_{i=1}^d y_i U(x_i) + U(W - C(x));$$

thus, letting $u(x) = -U(W - C(x))$ *and* $\Phi(x, y) = \delta \sum_{i=1}^d y_i U(x_i)$, *the optimal portfolio choice is therefore an element of* $\partial u^{\Phi}(y)$, *where*

$$\partial u^{\Phi}(y) = \arg\max_{x \in \mathcal{X}} \{\Phi(x, y) - u(x)\}.$$

Example 7.5 (Household labor supply with taxes). *Consider a household whose members have hourly wages* y_i ($i = 1, 2$). *Let* x_i *be the number of hours worked by each member. The total gross income of the household is* $x_1 y_1 + x_2 y_2$. *There is a convex tax schedule, thus the net income is* $\Phi(x, y) = F(x_1 y_1 + x_2 y_2)$, *where F is a concave function. Let* $d(x)$ *be the disutility (in monetary terms) associated with working* x_1 *and* x_2 *hours. The optimal number of hours worked is therefore an element of* $\partial d^{\Phi}(y) = \arg\max_{x \in \mathcal{X}} \{\Phi(x, y) - d(x)\}$.

Example 7.6 (Consumption choice with nonlinear budget sets). *Consider a consumer purchasing d goods, where the quantity of good i purchased is* x_i, *and the nominal price is* y_i. *Assume that there is a rebate* $r_i(x_i)$ *on the price of good i which is a function of* x_i, *so that the total cost of bundle x* is $c(x, y) = \sum_{i=1}^{d} (1 - r_i(x_i)) x_i y_i$. *Assume that the consumer's utility is quasi-linear in money and is given by $U(x) - c(x, y)$. Then, letting $\Phi(x, y) = -c(x, y)$ and $u(x) = -U(x)$, consumer x's consumption is an element of $\partial u^{\Phi}(y)$.*

7.2 THE MAIN RESULTS

We now state our first main result, which is an extension of theorem 2.2, with an additional condition on Φ so that a solution to the dual problem exists.

THEOREM 7.6 (Monge–Kantorovich duality again). *Let $\mathcal{X}$ and $\mathcal{Y}$ be two Banach spaces, and let P and Q be two probability measures on $\mathcal{X}$ and $\mathcal{Y}$ respectively. Let $\Phi : \mathcal{X} \times \mathcal{Y} \to \mathbb{R} \cup \{-\infty\}$ be an upper semicontinuous surplus function bounded from above and such that*

$$a_{\mathcal{X}}(x) + b_{\mathcal{Y}}(y) \leq \Phi(x, y) \leq c_{\mathcal{X}}(x) + d_{\mathcal{Y}}(y), \tag{7.5}$$

where $a_{\mathcal{X}}$ and $c_{\mathcal{X}}$ are integrable with respect to P, while $b_{\mathcal{Y}}$ and $d_{\mathcal{Y}}$ are integrable with respect to Q. Then, the value of the primal Monge–Kantorovich problem

$$\sup_{\pi \in \mathcal{M}(P,Q)} \mathbb{E}_{\pi} [\Phi(X, Y)] \tag{7.6}$$

coincides with the value of the dual

$$\inf_{\substack{u \in L^1(P), \\ v \in L^1(Q)}} \mathbb{E}_P [u(X)] + \mathbb{E}_Q [v(Y)] \tag{7.7}$$

$$\text{s.t. } u(x) + v(y) \geq \Phi(x, y),$$

and both problems have solutions.

A proof of this result can be found in [149, theorem 5.10]. Comparing with theorem 2.2 and the subsequent discussion, it appears that the condition needed for the existence of a solution to the dual problem is the first inequality in condition (7.5), while the condition needed for the existence of a solution to the primal problem is the second inequality in that condition. There are important special cases of condition (7.5):

- When $\Phi(x, y) = x'y$, we can take $a_{\mathcal{X}}(x) = -\frac{1}{2}|x|^2$ and $b_{\mathcal{Y}}(y) = -\frac{1}{2}|y|^2$; thus $a_{\mathcal{X}}$ and $b_{\mathcal{Y}}$ are integrable if P and Q have finite second-order moments, and theorem 6.3 appears as a consequence of theorem 7.6.
- When Φ is bounded from below and above respectively by l and L on $\mathcal{X} \times \mathcal{Y}$, one may take $a_{\mathcal{X}}(x) = b_{\mathcal{Y}}(y) = \frac{l}{2}$, and $c_{\mathcal{X}}(x) = d_{\mathcal{Y}}(y) = \frac{L}{2}$, and the assumptions of theorem 7.6 are met.

- When $\Phi(x, y) = -k(x - y)$ with k convex, as in example 7.2, one may take $a_{\mathcal{X}}(x) = -\frac{1}{2}k(2x)$, $b_{\mathcal{Y}}(y) = -\frac{1}{2}k(-2y)$, and the first inequality in condition (7.5) follows by the convexity inequality, while the other follows from the fact that Φ is then bounded from above.
- When $\Phi(x, y) = -d(x, y)$, where d is a distance, as in example 7.3, one may take $a_{\mathcal{X}}(x) = \Phi(x, z)$ and $b_{\mathcal{Y}}(y) = \Phi(y, z)$, for any z, and condition (7.5) follows by the triangle inequality. One needs to ensure the integrability assumptions on $a_{\mathcal{X}}$ and $b_{\mathcal{Y}}$ are met.

From an economic standpoint, condition (7.5) has an interesting interpretation. Indeed, we can interpret $a_{\mathcal{X}}(x)$ and $b_{\mathcal{Y}}(y)$ as the reservation utilities of types x and y being unmatched. With that interpretation, condition (7.5) implies that the *gain from matching*, which is the difference between the payoffs if matched and the sum of the payoffs if unmatched, namely, $\Phi(x, y) - a_{\mathcal{X}}(x) - b_{\mathcal{Y}}(y)$, is always nonnegative. In this case, as the reader is asked to verify in exercise 7.5 below, one may freely superimpose the constraints $u \geq a_{\mathcal{X}}$ and $v \geq b_{\mathcal{Y}}$ in problem (7.7), which means that there will be equilibrium payoffs which are always greater than or equal to the reservation utilities.

We now turn to the Monge problem.

THEOREM 7.7. *Let $\mathcal{X} = \mathbb{R}^d$ and $\mathcal{Y}$ be a Banach space. Let P and Q be two probability measures on $\mathcal{X}$ and $\mathcal{Y}$ respectively, such that P is continuous, and assume that*

(A1) *Φ is twice continuously differentiable and for every compact set $K \subseteq \mathcal{Y}$, there is $c_K > 0$ such that for every $x_1, x_2 \in \mathcal{X}$,*

$$\sup_{y \in K} |\Phi(x_1, y) - \Phi(x_2, y)| \leq c_K |x_1 - x_2|;$$

(A2) *the twist condition holds: for every $x \in \mathcal{X}$ and $y_1, y_2 \in \mathcal{Y}$,*

$$\nabla_x \Phi(x, y_1) = \nabla_x \Phi(x, y_2) \implies y_1 = y_2. \tag{7.8}$$

Then,

(i) *the value of the primal problem (7.6) is equal to the value of the dual problem (7.7), and both the primal and the dual problems have solutions;*

(ii) *the optimal coupling solution π to (7.6) is unique and is pure, so that there is a map $T : \mathcal{X} \to \mathcal{Y}$ such that π is the distribution of a random pair (X, Y) with*

$$Y = T(X), \quad X \sim P, \quad Y \sim Q; \tag{7.9}$$

hence T is a solution to the Monge problem;

(iii) *the solution (u, v) of the dual problem (7.7) is unique (up to an additive constant), and u is almost-everywhere differentiable;*

(iv) *T and u are related by*

$$T(x) = e_x(\nabla u(x)), \quad P\text{-}a.s., \tag{7.10}$$

where the map e_x is defined in (7.4).

The reader is referred to Carlier [31] for a proof.

7.3 COMPUTATION BY ENTROPIC REGULARIZATION

Although the discretization methods discussed in section 6.5 apply equally well to the case with general surplus function Φ, we conclude this section by discussing a slightly different class of methods called *entropic regularization*. These methods consist of slightly changing the formulation of the Monge–Kantorovich problem in order to make it easier to compute numerically, at least in approximation. More precisely, the objective function in the Monge–Kantorovich problem is augmented with an entropic term, so that the primal Monge–Kantorovich problem becomes

$$\max_{\pi \in \mathcal{M}(P,Q)} \mathbb{E}_\pi[\Phi(X, Y)] - \sigma \mathbb{E}_\pi[\ln \pi(X, Y)], \tag{7.11}$$

which, after defining f and g to be the densities of P and Q, has first-order solutions

$$\pi(x, y) = f(x)g(y)\exp\left(\frac{\Phi(x, y) - u(x) - v(y)}{\sigma}\right), \tag{7.12}$$

where the potentials u and v can be shown to be unique up to an additive constant. Letting π^σ be a solution to (7.11), and (u^σ, v^σ) be the corresponding potentials, one can show that when $\sigma \to 0$, π^σ and (u^σ, v^σ) converge to solutions to the primal and the dual problems, respectively. It turns out that when $\sigma \to +\infty$, $\pi^\sigma(x, y)$ converges to $f(x)g(y)$, which is the random coupling. This should come as no surprise: by consideration of problem (7.11), in the limit with very large σ, π tends to minimize $\mathbb{E}_\pi[\ln \pi(X, Y)]$ over $\pi \in \mathcal{M}(P, Q)$.

Note that the condition $\pi \in \mathcal{M}(P, Q)$ combined with (7.12) can be rewritten as

$$u(x) = \sigma \log \int_{\mathcal{Y}} g(y)\exp\left(\frac{\Phi(x, y) - v(y)}{\sigma}\right) dy, \tag{7.13}$$

$$v(y) = \sigma \log \int_{\mathcal{X}} f(x)\exp\left(\frac{\Phi(x, y) - u(x)}{\sigma}\right) dx. \tag{7.14}$$

This has several important consequences. First, when $\sigma \to 0$, relations (7.13) and (7.14) boil down to $u = v^{\Phi^\top}$ and $v = u^\Phi$:

REMARK 7.8 (Regularized Φ-conjugacy). Relations (7.13) and (7.14) provide a regularized version of the Φ-conjugacy relations $u = v^{\Phi^\top}$ and $v = u^\Phi$, toward which it converges when $\sigma \to 0$.

Second, relations (7.13) and (7.14) appear as the fixed point of an iterative procedure which consists of alternately expressing u from v and v from u. This suggests the following algorithm.

ALGORITHM 7.9 (Iterative proportional fitting procedure (IPFP)). *Take an initial guess of $v_0(y)$. At step $k \geq 0$, compute*

$$u_{k+1}(x) = \sigma \log \int_\mathcal{Y} g(y) \exp\left(\frac{\Phi(x, y) - v_k(y)}{\sigma} \right) dy,$$

$$v_{k+1}(y) = \sigma \log \int_\mathcal{X} f(x) \exp\left(\frac{\Phi(x, y) - u_{k+1}(x)}{\sigma} \right) dx.$$

Stop when v_{k+1} is sufficiently close to v_k.

This procedure enjoys many advantages: it is extremely fast, it is very simple to implement, and each of the updating steps (7.13) and (7.14) can be computed in parallel.

PROGRAMMING EXAMPLE 7.1. The program IPFP.R takes the same problem as programming example 3.1, and computes π via IPFP.

7.4 EXERCISES

Exercise 7.1 (M). **Distance cost.** *Prove the statements in example 7.3. [Hint: First show that $f^\Phi(y) - f^\Phi(x) \leq d(x, y).$]*

Exercise 7.2 (M). **Sum of generalized subdifferentials.** *Assume $\partial^\Phi f_1(x) = \{T_1(x)\}$ and $\partial^\Phi f_2(x) = \{T_2(x)\}$. Does one have $\partial^\Phi (f_1 + f_2)(x) = \{T_1(x) + T_2(x)\}$? [Hint: You may consider $\Phi(x, y) = \exp(x + y)$ and $f_1(x) = f_2(x) = \frac{1}{2}\exp(2x).$]*

Exercise 7.3 (C). **Parallelization.** *This exercise requires some notions in parallel computing. Using a parallel computation package in R, or using a version of MATLAB with parallel computing capabilities, modify the program in programming example 7.1 to take advantage of parallel computation.*

Exercise 7.4 (C). **Lowering the temperature.** *The aim of this exercise is to study the behavior of algorithm 7.9 when the temperature parameter tends to 0. In theory, $(\pi^\sigma, u^\sigma, v^\sigma)$ is supposed to tend to a solution to the*

Monge–Kantorovich problem as σ → 0. Take the program in programming example 7.1, and experiment with lower values of σ. Below which value of σ do you start seeing problems? Suggest a fix for this problem.

Exercise 7.5 (E). ***Reservation values****. Under the assumptions of theorem 7.6, show that the value of the dual problem (7.7) is unmodified if the additional constraints $u(x) \geq a_{\mathcal{X}}(x)$ for all $x \in \mathcal{X}$ and $v(y) \geq b_{\mathcal{Y}}(y)$ for all $y \in \mathcal{Y}$ are superimposed.*

Exercise 7.6 (E). ***A multiplicative model****. Consider a labor market where firms are endowed with a job of complexity $y \in \mathcal{Y}$ drawn from a continuous distribution Q. Workers are endowed with skills (ability) $x \in \mathcal{X}$ drawn from a continuous distribution P. To produce one unit of output at job y, a worker x must work for $K(x, y)$ units of time. Let $W(x)$ be the wage function indicating the per-unit-of- time wage for worker x (endogenously determined). The total cost for a firm y of employing worker x is then given as $K(x, y)W(x)$. It is assumed that the output is not differentiated within workers. Show that this problem can be reformulated as a Monge–Kantorovich problem.*

7.5 REFERENCES AND NOTES

The main reference for this chapter is Villani [149, chapter 5], complemented by Villani [148], and the papers by Carlier [29, 30, 31]. Our exposition on generalized convexity follows [149, chapter 5] (with different conventions for signs and notation). Example 7.2 is studied in [71]. Example 7.3 constituted the Kantorovich–Rubinstein theorem, which appeared in a 1958 paper [85]. The reference for theorem 7.6 is [149, theorem 5.10]. The reference for theorem 7.7 is Carlier [31], which is sufficient for many purposes; there are more general variants in [149, chapter 10], but some conditions there are much more technical. In section 7.3, equation (7.12) is sometimes called the *Schrödinger–Bernstein* equation, studied, for example, in [130]; see also a survey in [94]. Zambrini discovered a connection between a dynamic version of this equation and Ed Nelson's stochastic mechanics; see [154]. It has been applied to economics in [68, 69, 47]. Algorithm 7.9 is old and has been rediscovered many times. An early reference is Deming and Stephan [45]; see convergence results in [127]. Exercise 7.6 is inspired by Sattinger [133].

8

Transportation on Networks

Our focus so far when we have considered optimal transport has been the cost of going from A to B, not really the way we went from A to B. In some cases, this is a trivial problem, as in the fountains problem, where individuals may go to fountains along a straight line; however there may be situations where this assumption is not realistic (for example, urban transportation is done through a network). Hence, we shall focus more deeply on transportation on networks, when one may go from A to B through a well-defined network only, incurring some well-defined cost to move from a node of the network to an adjacent one.

As a motivating example, let us think of the ancient Silk Roads. The secret of silk manufacturing was one of the best-kept secrets of ancient times, and remained unknown outside China for thousands of years. Europeans immensely valued the fabric, which was sold at sky-high prices in Rome and Byzantium. The silk was shipped via a network of routes through Central Asia and by the sea, with several production and consumption locations, and a number of possible itineraries. Figure 8.1 is a sketch of these routes. The goods were shipped on caravans, which did not go along the whole route, but instead went back and forth over distances which rarely exceeded 200 or 300 miles. Silk was therefore shipped steadily from China to the Roman Empire in spite of the virtual absence of contact between these far apart civilizations. A powerful force moved the precious fabric over 6000 miles, across deserts, mountains, all sorts of kingdoms and tribal areas, more surely than any planned expedition; and unambiguous signals showed the way to Byzantium better than any compass. These signals, of course, are prices: the price of silk (in its gold equivalent) steadily increased between Xi'an or Guangzhou all the way to Antioch or Rome; and the prospect of obtaining a portion of the price increase led traders to venture in caravans across the steppes.

In this chapter, we shall model the problem of shipping a good produced and consumed at various locations, through a network. As before, we shall first study the problem of optimality, which is the central planner's problem: how should all of the silk be shipped from the production locations to the consumption locations at minimal cost? We will then move on to the equilibrium problem: is there a system of prices for the silk at any point

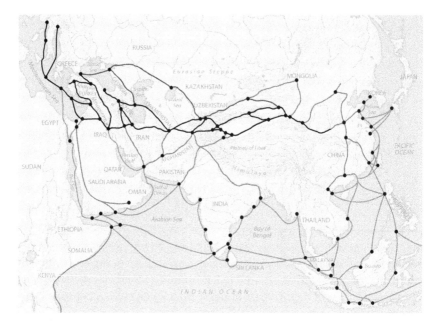

Figure 8.1: The Silk Roads.

in the network such that the combined actions of each trader, acting in a decentralized way, lead to shipping all of the silk from production locations to consumption locations? As before, we shall see that in our setting, the answers to these two questions are in fact equivalent, and that a duality result is underlying.

While this book is about optimal transportation, this chapter is the only chapter where we will describe literally a transportation problem: the question addressed in this chapter is not only which production locations produce for which consumption locations (which is an optimal assignment problem), but also, given a pair of production and consumption locations, what is the path through which to ship the silk in an optimal way (which is an optimal path problem). Hence, the formalism we shall introduce in this chapter embeds optimal assignment problems and optimal path problems. In order to achieve this goal, we need to have a formal description of the *network* on which the silk can be shipped. This is what we set out to do next.

8.1 SETUP

We start by introducing the basic formalism needed to describe a transportation network. Essentially, the exogenous quantities we need to specify are

- the topology of the network: which pairs of cities are directly connected;
- a metric on the network: the surplus (or cost) of shipping the good between two directly connected cities;
- supply and demand: which cities produce, which cities consume, and which cities are just trading intermediaries.

We will then need to determine the outcome, which consists of

- the network flow, which is the quantity of good shipped between each pair of directly connected cities;
- the potentials, which have an interpretation in terms of prices of the good in any city in the decentralized solution.

8.1.1 The Network

8.1.1.1 Topology

We start by defining the directed graph on which transportation takes place.

DEFINITION 8.1 (Directed graph). *A (directed) graph $(\mathcal{Z}, \mathcal{A})$ is a set of nodes (cities) $\mathcal{Z}$, along with a set of arcs $\mathcal{A} \subseteq \mathcal{Z}^2$ which are pairs (x, y) where $x, y \in \mathcal{Z}$.*

Note that these are *directed* arcs, and thus an (x, y) arc is distinct from a (y, x) arc. If $(x, y) \in \mathcal{A}$, then we say that x is *directly connected* to y. Unless explicitly stated otherwise, we will assume that a node is never directly connected to itself.

Very often, we have to compare quantities (usually prices) between two directly connected cities x and y. If w_x is the price of silk at city x, and if there is an arc from x to y, then $w_y - w_x$ is the price gradient along the (x, y) arc. This is expressed using a matrix as follows.

DEFINITION 8.2 (Gradient matrix). *We define the* gradient matrix *(also called an* edge-node matrix*) as the matrix with general term ∇_{ax}, $a \in \mathcal{A}$, $x \in \mathcal{Z}$, such that*

$$\nabla_{ax} = \begin{cases} -1 & \text{if } a \text{ is out of } x, \\ +1 & \text{if } a \text{ is into } x, \\ 0 & \text{otherwise.} \end{cases}$$

Hence, for a vector $(w_x)_{x \in \mathcal{Z}}$, the *gradient* of w at arc $(x, y) \in \mathcal{A}$ is $(\nabla w)_{xy} = w_y - w_x$, which is defined only if $(x, y) \in \mathcal{A}$, that is, if there is an arc from x to y. Note that our notation and terminology are chosen to stress the analogy with the corresponding differential operators in the continuous case.[1]

---

[1] This analogy is very fruitful and can be pushed much further; see an interesting account in Grady and Polimeni [72].

Finally, we define two important concepts: paths and loops.

DEFINITION 8.3 (Path). *Given two nodes x and y, a* path *from x to y is a sequence $x_1, x_2, \ldots, x_K$ in $\mathcal{Z}$, where $x_1 = x$, $x_K = y$, and for every $1 \le k \le K - 1$, $(x_k, x_{k+1}) \in \mathcal{A}$.*

DEFINITION 8.4 (Loop). *A* loop *(also called a* cycle*) is a path from a node x to itself.*

8.1.1.2 Transportation Surplus

We shall assume that if node x is directly connected to node y, then the *transportation surplus* of transporting a unit of mass through (x, y) is given by Φ_{xy}. Therefore, Φ is a vector defined on $\mathbb{R}^{\mathcal{A}}$. Equivalently, $C_{xy} = -\Phi_{xy}$ may be thought of as a *transportation cost*, reflecting distance, discomfort, risk, transit time, etc. The surplus of moving the good from node x to node y along path $x_1, x_2, \ldots, x_K$ is

$$\sum_{k=1}^{K-1} \Phi_{x_k x_{k+1}}.$$

A natural and important assumption is that moving the good along a loop does not yield a positive surplus. This is consistent with our physical intuition that there is no perpetual motion: transportation has a cost. If we did not make this assumption, the solutions would involve an indeterminate quantity of such artificial loops.

ASSUMPTION 8.5 (No profitable loop). *There is no profitable loop, which means that there is no sequence $x_1, \ldots, x_K$ in $\mathcal{Z}$ such that $x_K = x_1$, $(x_k, x_{k+1}) \in \mathcal{A}$, and $\sum_{k=1}^{K-1} \Phi_{x_k x_{k+1}} > 0$.*

In particular, there is no profitable loop if $\Phi \le 0$.

8.1.1.3 Supply and Demand

For each $x \in \mathcal{Z}$, let n_x be the *net demand*, which is the flow of goods disappearing from the graph. If $n_x < 0$, there is an actual supply of goods at node x, which is then called a *supply node* (often called a *source node* in the literature on networks), which means that a quantity $|n_x|$ of the good is produced at location x. If $n_x > 0$ there is an actual demand for goods at node x which is called a *demand node* (often called a *terminal node*), which means that a quantity n_x of the good is consumed at location x. The set of supply and demand nodes are denoted $\mathcal{X}$ and $\mathcal{Y}$, respectively; thus

$$\mathcal{X} = \{x \in \mathcal{Z} : n_x < 0\} \quad \text{and} \quad \mathcal{Y} = \{x \in \mathcal{Z} : n_x > 0\}.$$

We shall make two important assumptions about supply and demand. First, we shall assume that total supply equals total demand on the network. This assumption is necessary as we will not allow free disposal, thus the conservation of mass equation discussed below will imply that the total mass that appears on the network should disappear in the same quantity. Total supply is $-\sum_{x\in\mathcal{X}} n_x$, total demand is $\sum_{y\in\mathcal{Y}} n_y$. Thus equality between total supply and total demand can be rewritten as $\sum_{x\in\mathcal{X}} n_x + \sum_{y\in\mathcal{Y}} n_y = 0$.

ASSUMPTION 8.6 (Balancedness). *Assume that total supply equals total demand on the network, that is,*

$$\sum_{x\in\mathcal{X}} n_x + \sum_{y\in\mathcal{Y}} n_y = 0.$$

Second, we need to rule out impossibilities that may arise, for instance if there is no path connecting any supply node to any demand node. While there are interesting intermediate cases, we will simply make the strong assumption that any demand node can be reached from any supply node via a path on the network.

ASSUMPTION 8.7 (Connectedness). *Assume the set of supply nodes $\mathcal{X}$ is strongly connected to the set of demand nodes $\mathcal{Y}$, that is, for every $x \in \mathcal{X}$ and $y \in \mathcal{Y}$, there is a path from x to y.*

8.1.1.4 Summary: Network

The specification of the graph, the net demand vector and the surplus vector, defines a network.

DEFINITION 8.8 (Network). *A directed graph $(\mathcal{Z}, \mathcal{A})$, endowed with a net demand vector $(n_z)_{z\in\mathcal{Z}}$ and a surplus vector $(\Phi_a)_{a\in\mathcal{A}}$, is called a* network $(\mathcal{Z}, \mathcal{A}, n, \Phi)$.
If assumptions 8.5 (no profitable loop), 8.6 (total supply equals total demand), and 8.7 (supply is strongly connected to demand) all hold, the network is called regular.

8.1.2 Outcome

We now describe outcomes, which are the endogenous quantities. In the bipartite case, the outcome consists of (π, u, v), where the matching distribution (π_{xy}) is a solution to the primal Monge–Kantorovich problem, and where the equilibrium payoffs (u_x, v_y) are a solution to the dual problem, which can be interpreted as prices. Similarly here, an outcome will consist of a vector (π, v) where (π_{xy}) is a vector defined on every arc (x, y) which is the flow of goods through that arc, and (v_x) is a vector defined on every node x, which is interpreted as (minus) the price of the good at node x.

8.1.2.1 Network Flow

For $(x, y) \in \mathcal{A}$, let $\pi_{xy} \geq 0$ be the *flow* of goods through arc (x, y), which is interpreted as the quantity of silk that is shipped directly from city x to city y. The *conservation constraints* require that the total intermediate flow into a node minus the total intermediate flow out of a node must equal the exiting flow. This means that the sum of the silk produced and the silk arriving in a city equals the sum of the silk consumed and the silk leaving; namely,

$$\sum_{z \in \mathcal{Z}} \pi_{zx} - \sum_{z \in \mathcal{Z}} \pi_{xz} = n_x \tag{8.1}$$

should hold for all x. This equation is sometimes also called *Kirchhoff's law*. Clearly, constraint (8.1) is linear with respect to π. In the spirit of the discussion in section 3.4, we would like to express this constraint in matrix form. This will be done thanks to the following result.

DEFINITION 8.9 (Incidence matrix). *The* incidence matrix *(sometimes also called the* node-edge matrix*) is the transpose of the gradient matrix, and therefore denoted* ∇^*.

Thus $\nabla^*_{xa} := \nabla_{ax}$. As a result, for a flow π,

$$\left(\nabla^* \pi \right)_x = \sum_{z:(z,x) \in \mathcal{A}} \pi_{zx} - \sum_{z:(x,z) \in \mathcal{A}} \pi_{xz},$$

and conservation equation (8.1) can be rewritten as $\nabla^* \pi = n$. This motivates the following definition.

DEFINITION 8.10 (Feasible flows). *The set of feasible flows, denoted* $\mathcal{M}(n)$, *or* $\mathcal{M}$ *when there is no ambiguity, is defined as the set of flows* $\pi \geq 0$ *that verify the conservation equation*

$$\nabla^* \pi = n. \tag{8.2}$$

We shall often omit the adjective "feasible," implicitly meaning feasible flows when mentioning only flows. Clearly, the notation $\mathcal{M}(n)$ is purposely chosen in order to resemble the notation $\mathcal{M}$ in previous chapters.

8.1.2.2 Prices and Potentials

Finally, we assume that a price for silk is defined at every city $x \in \mathcal{Z}$. Let w_x be this price. Note that if x is directly connected to y, then $w_y - w_x + \Phi_{xy}$ is the profit of the strategy that consists of buying silk at city x, shipping it to y, and selling it at y; at equilibrium, this cost cannot be positive, otherwise it would yield an arbitrage opportunity. Hence, $w_y - w_x + \Phi_{xy} \leq 0$. In matrix terms, this can be expressed as $\nabla w + \Phi \leq 0$, or, letting $v = -w$,

$$\nabla v \geq \Phi. \tag{8.3}$$

8.2 OPTIMAL FLOW PROBLEM

The surplus of shipping π_{xy} units of silk from city x to city y is assumed to be proportional to the unit surplus, and thus it is equal to $\pi_{xy}\Phi_{xy}$. This implies that the marginal transportation cost is constant, which implicitly assumes away congestion (which would lead to an increasing marginal cost), or fixed costs or economies of scale (which would lead to a decreasing marginal cost). The total surplus (minus the total cost) associated with network flow π is thus $\sum_{(x,y)\in\mathcal{A}} \pi_{xy}\Phi_{xy}$. Assuming that all operations are managed by a central planner, the problem then becomes to maximize the total surplus of transportation under the feasibility constraint, and hence solve the *optimal flow problem*

$$\max_{\pi\geq 0} \sum_{(x,y)\in\mathcal{A}} \pi_{xy}\Phi_{xy} \tag{8.4}$$

$$\text{s.t. } \nabla^*\pi = n,$$

which is a linear programming problem whose dual is

$$\min_{v\in\mathbb{R}^{\mathcal{Z}}} \sum_{x\in\mathcal{Z}} n_x v_x \tag{8.5}$$

$$\text{s.t. } \nabla v \geq \Phi.$$

However, unlike the bipartite case seen in chapter 3, the values of these problems are not automatically finite, and we need to impose assumptions 8.5 and 8.7 to ensure that they are, by virtue of the following result, whose proof is left as an exercise.

PROPOSITION 8.11.   (i) *Under assumption 8.5 (no profitable loop), the dual problem (8.5) is feasible, which means that there is a vector $v \in \mathbb{R}^{\mathcal{Z}}$ such that $\nabla v \geq \Phi$; and the value of problem (8.4) is strictly less than $+\infty$.*

(ii) *Under assumptions 8.6 (total supply equals total demand) and 8.7 (supply is strongly connected to demand), the primal problem (8.4) is feasible, which means that there is a flow $\pi \geq 0$ such that $\nabla^*\pi = n$; and the value of problem (8.5) is strictly greater than $-\infty$.*

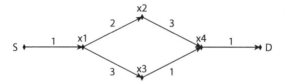

Figure 8.2: A network with one supply node (S) and one demand node (D).

Throughout the remainder of the chapter we will hence maintain assumptions 8.5, 8.6, and 8.7. In other words, we shall assume that $(\mathcal{Z}, \mathcal{A}, n, \Phi)$ is a "regular" network in the sense of definition 8.8.

The optimal flow problem is a linear programming problem. We now give the fundamental duality result which generalizes theorem 3.1. The proof directly follows from standard linear programming duality and is left as an exercise.

THEOREM 8.12 (Linear programming duality). *Assume that $(\mathcal{Z}, \mathcal{A}, n, \Phi)$ is a regular network. Then the value of the primal problem (8.4) coincides with the value of its dual (8.5), and both problems have solutions. Further, if π is a solution to the primal and v is a solution to the dual, then $\pi_{xy} > 0$ implies $v_y - v_x = \Phi_{xy}$.*

To give an economic interpretation of this theorem, let us go back to the discussion started in section 8.1.2.2. Recall that if v is a solution to the dual program, then $w_x = -v_x$ can be interpreted as the price of the good at location x. We have argued that $w_y - w_x + \Phi_{xy}$, which is the profit of the strategy that consists of buying silk at city x, shipping it to y, and selling it at y, cannot be positive at equilibrium, otherwise it would yield an arbitrage opportunity. Complementary slackness tells us that along the arcs where there is a positive flow of the good, the profit of this strategy is 0. Otherwise, the prices could not be sustained.

Note that if v is a solution to the dual problem and $c \in \mathbb{R}$, then $v + c$ is still a solution to the dual problem. As before, this indeterminacy comes from the fact that we have not introduced an outside option, which would provide an upper bound for v in supply nodes (i.e., a lower bound for prices in supply nodes), and a lower bound in demand nodes (i.e., an upper bound for prices in demand nodes).

We illustrate theorem 8.12 with two examples. The first example is a simple optimal flow problem with one supply node and one demand node.

Example 8.1. *Consider the network described in figure 8.2, with $\mathcal{Z} = \{S, D, x_1, \ldots, x_4\}$ where Φ is apparent on the figure, and the net supply vector is such that $n_S = -1$, $n_D = 1$, and $n_{x_i} = 0$ for $i = 1, \ldots, 4$. The optimal flow has value 7 and it gives no flow on arcs (x_1, x_3) and (x_3, x_4), and the value of the flow equals 1 on all the other arcs. As a consequence, if the dual potential is,* *for example, normalized by* $v_S = 0$, *then* $v_{x_1} = 1$, $v_{x_2} = 3$, $v_{x_4} = 6$, *and* $v_D = 7$. *The value of* v_{x_3} *is constrained by inequalities* $v_{x_3} - v_{x_1} \geq \Phi_{x_1 x_3}$ *and* $v_{x_4} - v_{x_3} \geq \Phi_{x_3 x_4}$; *thus*

$$5 = v_{x_4} - \Phi_{x_3 x_4} \geq v_{x_3} \geq v_{x_1} + \Phi_{x_1 x_3} = 4.$$

One recovers that the value of the primal problem $\sum_{a \in \mathcal{A}} \pi_a \Phi_a$ *coincides with the value of the dual, which is* $n_S v_S + n_D v_D = 7$.

Note that in example 8.1, one unit of good is supplied to the network at the supply node S, and one unit is consumed at the demand node D. More generally, if one thinks of $l_a = -\Phi_a$ as the length of arc a, the problem becomes a shortest path problem, as we argue in our second example.

Example 8.2 (Shortest path). *Consider a regular network* $(\mathcal{Z}, \mathcal{A}, n, \Phi)$ *and assume there is one supply node* S *and one demand node* D *such that* $n_z = 1$ $\{z = D\} - 1\{z = S\}$ *for all* $z \in \mathcal{Z}$. *Assume* $\Phi_a = -l_a$, *where* l_a *is the metric length of arc* a. *Then, letting* (π, v) *be a pair of solutions to the primal and dual optimal flow problems, the shortest path from* S *to* D *is given by* $v_S - v_D$, *while a shortest path is constructed recursively by the following algorithm:*

- *At initial step* $k = 0$, *take* $z_0 = S$.
- *At step* k, *if* $z_k = D$ *then stop. Else, by conservation of mass, there is some* $z' \in \mathcal{Z}$ *such that* $z_k z' \in \mathcal{A}$ *and* $\pi_{z_k z'} > 0$. *Then choose* $z_{k+1} := z'$ *and move to step* $k + 1$.

This example, including the algorithm, is implemented in programming example 8.1.

In our third example, we show that the setting of the optimal assignment problem seen in chapter 3 can be recast as a particular instance of an optimal flow problem on a network.

Example 8.3 (Bipartite assignment). *Let us go back to the case considered in chapter 3, and let us show that this is a particular case of the setting considered in the present chapter. In this case, supply nodes are the set of workers, and demand nodes are the set of firms. Let the set of nodes* $\mathcal{Z} = \mathcal{X} \cup \mathcal{Y}$, *while the arcs connect each firm to each worker, thus* $\mathcal{A} = \mathcal{X} \times \mathcal{Y}$. *The definition of* Φ_{xy} *is transparent. Excess demand is given by* $n_x = -p_x$ *if* $x \in \mathcal{X}$, *and* $n_y = q_y$ *if* $y \in \mathcal{Y}$. *This is called a* bipartite network. *In this type of network, paths are made of single arcs* (x, y) *from* $x \in \mathcal{X}$ *to* $y \in \mathcal{Y}$, *and there are no loops. The definition of* π_{xy} *is transparent too, and the conservation equation* $\nabla^* \pi = n$ *is rewritten as*

$$\sum_{y \in \mathcal{Y}} \pi_{xy} = p_x \ \forall x \in \mathcal{X} \quad \text{and} \quad \sum_{x \in \mathcal{X}} \pi_{xy} = q_y \ \forall y \in \mathcal{Y}.$$

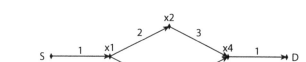

Figure 8.3: A simple modification of the network from figure 8.2, that has multiple
optimal flows.

The dual variable v_x coincides with what we previously denoted $-u_x$ for
$x \in \mathcal{X}$, while, for $y \in \mathcal{Y}$, it coincides with what we previously denoted v_y.
Thus the objective function of the dual problem is $\sum_{x \in \mathcal{X}} p_x u_x + \sum_{y \in \mathcal{Y}} q_y v_y$,
while the constraints become $u_x + v_y \geq \Phi_{xy}$. Hence theorem 8.12 boils down to
theorem 3.1 in the case when the network is bipartite.

8.3 INTEGRALITY

Extending the argument in section 3.3, we will now discuss integrality.
Assuming n_x is a positive or negative integer for all $x \in \mathcal{Z}$, the optimal flow π
need not be integral. In example 8.1 and the network in figure 8.2, the optimal
flow π is indeed integral as it is valued in $\{0, 1\}$; however, a simple modification
of that network has nonintegral optimal flows, as seen in example 8.4.

Example 8.4. *Consider the network in figure 8.3, which is almost identical
to the one in figure 8.2, with the only difference being that the value of $\Phi_{x_1 x_3}$
is 4 instead of 3. As a result, both the upper path $x_1 x_2 x_4$ and the lower path
$x_1 x_3 x_4$ have length 5. While this does not change the value of the problem,
which is still 7, this implies that the optimal flow can split at node x_1 and send
a fraction $\theta \in [0, 1]$ of the mass in the upper path, and a fraction $(1 - \theta)$ of the
mass in the lower path. In this case, the corresponding optimal flow π will have
$\pi_{x_1 x_2} = \pi_{x_2 x_4} = \theta$ and $\pi_{x_1 x_3} = \pi_{x_3 x_4} = 1 - \theta$. Therefore, π is an integral solution
if and only if $\theta \in \{0, 1\}$.*

It is however desirable to find integral flows. Indeed, it is not hard to see
that integral feasible flows can be represented as a superposition of paths,
as the reader is asked to verify in exercise 8.2. Despite the existence of
nonintegral solutions in example 8.4, it is clear that the problem considered in
that example has integral solutions, one of which consists of sending all the
flow through the upper path, and one of which consists of using the lower
path. Theorem 8.13 below argues that this holds quite generally: as soon as
n_x is integral, there are integral solutions.

THEOREM 8.13 (Integral flow theorem). *Consider a regular network* $(\mathcal{Z}, \mathcal{A}, n, \Phi)$ *and assume that n_z is a positive or negative integer for all $z \in \mathcal{Z}$. Then there is a solution π to problem (8.4) such that π_z is an integer for all $z \in \mathcal{Z}$.*

PROOF. The proof builds on the same ideas as the proof of theorem 3.4. Consider a solution π to problem (8.4), and let p be the number of nonintegral entries in π . We will show that if $p > 1$, then there is another solution to problem (8.4) with at most $p - 1$ nonintegral entries. Indeed, start from a nonintegral entry $\pi_{x_0 x_1}$. Then one can show that one can build a sequence of nodes $x_0, x_1, \ldots, x_t$ such that $x_t = x_i$ for some $0 \leq i \leq t$; hence $x_i, x_{i+1}, \ldots, x_t$ is a loop. Let $\delta\pi$ be such that $\delta\pi_{x_{i+j}x_{i+j+1}} = (-1)^j$, $\delta\pi = 0$ otherwise; $\delta\pi$ is the flow associated with what is called an *alternating path*. One has $\nabla^* \delta\pi = 0$. Let $\bar{\epsilon} > 0$ be the largest value of ϵ such that $\pi + \epsilon\,\delta\pi \geq 0$. Letting $\pi' = \pi + \bar{\epsilon}\delta\pi$, it is easy to see that π' is still feasible, optimal, and that the number of its nonintegral entries is at most $p - 1$. By induction, there is an integral solution to problem (8.4). □

Theorem 8.13 is very useful because it implies that the shortest path problem can be solved using linear programming methods, as will be demonstrated in programming example 8.1, where the shortest path in the network of the Parisian subway is determined using the linear programming formulation. While the optimal π returned by the solver is not necessarily integral, the very simple algorithm described in example 8.2 finds a shortest path based on an optimal π (not necessarily integral). In the particular case of the optimal assignment problem, it is not hard to see that this result implies theorem 3.4.

8.4 COMPUTATION VIA LINEAR PROGRAMMING

There are very fast algorithms to compute optimal flow problems using their intrinsic structure. Actually most algorithms that exist for the optimal assignment problem have a more general version for optimal flow problems; this is, in particular, the case for the Hungarian algorithm, Bertsekas' auction algorithm, and many others. Many of these algorithms are described in [1] and [18]. Here, as in chapter 3, we shall limit ourselves to showing how to compute the optimal assignment using a general purpose linear programming solver such as Gurobi.

Consider a network with nbNodes nodes and nbArcs arcs. The typical structure of the data needed is

(i) arcs, an array of integers of size nbArcs×2 whose entries are between 1 and nbNodes, so that each line describes an arc, represented by its origin node (first column) and its destination node (second column);

(ii) n, a vector of length nbNodes such that n[x] is the net demand at node x;

(iii) Phi, a vector of length nbArcs such that Phi[a] is the transportation surplus at arc a.

The node-incidence matrix is coded in a sparse way. In R, this is done using the package Matrix [10], and the relevant instruction is

```
Nabla =
sparseMatrix(i=1:nbArcs,j=arcs[,1],dims=c(nbArcs,nbNodes),x=-1)
+sparseMatrix(i=1:nbArcs,j=arcs[,2],dims=c(nbArcs,nbNodes),x=1).
```

The model is set up for Gurobi as

```
sol = gurobi(list(A=t(Nabla),obj=Phi,modelsense='max',
rhs=n,sense='=')).
```

We run Gurobi. The optimal flow pi and the potential v are computed using

```
pi = sol$x
v = sol$pi.
```

PROGRAMMING EXAMPLE 8.1. In the program shortestPath.R, we compute your optimal path in the Parisian subway. More precisely, we determine the path that minimizes the total geographic distance traveled. Figure 8.4 shows a map of the Paris subway. The network is encoded[2] in the file arcs.csv, which describes the arcs: each line corresponds to an arc in the network. It has three columns. The first column contains the index of the origin station; the second column contains the index of the destination station; and the third column is the geographic distance between origin and destination. The first line of this file reads

1   2   1027.280873

meaning that there is an arc of length 1027.28 meters going from station 1 to station 2. In the file names.csv, each arc is associated with the name of the station. The first two lines of this file read

La Défense (Grande Arche)
Esplanade de la Défense

meaning that arc 1's name is La Défense (Grande Arche) and arc 2's name is Esplanade de la Défense. Nabla is set as described above; Phi is set as minus the third column of arcs.csv. Assume that one would like to go from station #84 (Saint-Germain des Prés) to station #116 (Trocadéro). The net demand vector is initialized as a vector of 0s, of length nbNodes, and

---

[2] The network data are available on www.ratp.fr/opendata. Thanks to Lucas Vernet for his help with preparing the data in their current format.

Figure 8.4: A map of the Paris subway.

set as follows: `n[c(84,116)] = c(-1,1)`. Finally, the simple algorithm described in example 8.2 determines an optimal path from an optimal solution π returned by the solver.

8.5 EXERCISES

Exercise 8.1 (M). *Gradient flows. Consider a network $(\mathcal{Z}, \mathcal{A}, n, \Phi)$.*

   (i) Show that the network has no profitable loop if and only if there exists a potential (u_x) such that

$$\forall (x, y) \in \mathcal{A}, \quad u_y - u_x \geq \Phi_{xy}. \tag{8.6}$$

(ii) *Show that this implies* $\Phi_{xy} + \Phi_{yx} \leq 0$ *whenever* (x, y) *and* (y, x) *are in* $\mathcal{A}$.

(iii) *Assume that* $(x, y) \in \mathcal{A}$ *implies* $(y, x) \in \mathcal{A}$ *and assume* Φ *is antisymmetric, that is,* $\Phi_{xy} = -\Phi_{yx}$ *for every* $(x, y) \in \mathcal{A}$. *Show that condition (8.6) implies* $\Phi_{xy} = u_y - u_x$.

Exercise 8.2 (M). **Paths and loops.** *Consider a regular network* $(\mathcal{Z}, \mathcal{A}, n, \Phi)$, *and let* $\pi \in \mathcal{M}(n)$ *be a feasible flow. Assume that* π *is integral, meaning that all* $\pi_{xy}, xy \in \mathcal{A}$ *are integers.*

(i) *Show that* π *can be represented as a superposition of paths from supply nodes to demand nodes, and loops.*

(ii) *Assume that* π *is optimal for problem (8.4). What further restriction is imposed on the previous representation?*

Exercise 8.3 (E). **A dynamic network.** *Show how to construct a dynamic version of the optimal flow problem described in this chapter, where the net supply n and transportation surplus* Φ_a *not only depend on x and a respectively, but also on time* $t \in \{1, \ldots, T\}$. *Argue why this extension is important.*

Exercise 8.4 (E). **A network with free disposal.** *Consider a variant of problem (8.4) where if* $x \in \mathcal{X}$ *is a supply node, then the quantity entering the network at node x is at most* $|n_x|$, *and if* $y \in \mathcal{Y}$ *is a demand node, then the quantity leaving the network at node y is at most* n_y. *Write down the correct constraints for a suitably modified primal problem, and write down the dual problem.*

Exercise 8.5 (C). **Computations on a dynamic network.** *Implement the dynamic version of the optimal flow problem from exercise 8.3, based on the code provided in programming example 8.1.*

Exercise 8.6 (C). **Computations on a network with free disposal.** *Implement the linear program from exercise 8.4, based on the code provided in programming example 8.1.*

8.6 REFERENCES AND NOTES

There are many references on network flow problems. We refer in particular to Rockafellar [123] for an analytic treatment, and Ahuja et al. [1] or Bertsekas [18] for more computational aspects. A nice, condensed treatment is given in Vohra [151, chapter 3], which gives applications to mechanism design. The set of lecture notes [61] has more on congestion problems, which are not covered here. A reference for exercise 8.3 is [122, section 1H].

9

Some Applications

This chapter presents a number of applications of optimal transport theory to economics. As will soon become apparent, these applications are biased toward the author's own research. The emphasis is put on the formal structures, in particular in connection with optimal transport. Thus the results are presented somewhat loosely, and empirical applications are not discussed at all. The reader is referred to the original papers for further detail.

9.1 RANDOM SETS AND PARTIAL IDENTIFICATION

9.1.1 The Problem

In this section, we shall deal with problems of partial identification in econometrics. Partial identification may arise when observations are missing or incomplete (which is the case with censoring, or bracketing), or when models themselves are incompletely specified (this is the case when models predict multiple equilibria and an equilibrium selection mechanism is not chosen). To make things more concrete in a very stylized example, assume that the econometrician observes the distribution P of some incomplete measure of economic outcome $X \in \mathcal{X}$. Measurement $x \in \mathcal{X}$ is incomplete in the sense that the set of outcomes $y \in \mathcal{Y}$ compatible with observation x is a closed set $\Gamma_\theta(x) \subseteq \mathcal{Y}$, where $\theta \in \Theta$ is a parameter of the model. Here $\Gamma_\theta : \mathcal{X} \to 2^{\mathcal{Y}}$ is a *set-valued function*, also called a *correspondence*. For instance, observing an individual's marginal tax rate x is informative about this individual's income bracket, but not about the precise value of this income. Finally, assume that a model (also parameterized by $\theta \in \Theta$) predicts that the actual outcome Y should follow a distribution Q_θ over $\mathcal{Y}$.

The model is compatible with the measurement when there is a coupling (X, Y), where the observed measurement X is distributed as P, the latent outcome Y is distributed as Q_θ, and $Y \in \Gamma_\theta(X)$ holds almost surely. Letting $G_\theta(y)$ be the set of y such that $y \in \Gamma_\theta(x)$, G_θ is called the inverse correspondence to Γ_θ, and the problem can be equivalently reformulated in order to check that $X \in G_\theta(Y)$.

DEFINITION 9.1. *The identified set $\Theta_I \subseteq \Theta$ is the set of parameters $\theta \in \Theta$ such that there is a coupling $(X, Y) \sim \pi \in \mathcal{M}(P, Q_\theta)$ for which almost surely,*

$$X \in G_\theta(Y).$$

The problem is therefore the following.

PROBLEM 9.2. *Given a value $\theta \in \Theta$, how do we decide whether $\theta \in \Theta_I$ or not?*

As we shall see, this problem can be reformulated as a Monge–Kantorovich problem. This reformulation will be very useful for computational purposes.

9.1.2 Random Sets and Optimal Transport

Note that $\mathcal{G} := G_\theta(Y)$ is a random set, more precisely a random closed set. A very important notion with random closed sets is their *capacity*, which as we shall see actually characterizes the distribution.

DEFINITION 9.3. *The capacity associated with the random set $\mathcal{G}$ is the function $c_\mathcal{G}$ defined on measurable sets such that*

$$c_\mathcal{G}(B) := \Pr(\mathcal{G} \cap B \neq \emptyset).$$

Note that, in the case when $\mathcal{G}$ is a singleton containing a random variable Z, that is, when $\mathcal{G} = \{Z\}$, the associated capacity is simply $c_{\{Z\}}(B) = \Pr(Z \in B)$, which is the distribution function associated with Z. For this reason, $c_\mathcal{G}$ is sometimes simply called the distribution function associated with the random set $\mathcal{G}$. However, as soon as $\mathcal{G}$ is multivalued, its distribution has an important difference compared to the distributions of random variables: it is not additive, in the sense that $B_1 \cap B_2 = \emptyset$ does not necessarily imply $c_\mathcal{G}(B_1 \cup B_2) = c_\mathcal{G}(B_2) + c_\mathcal{G}(B_1)$.

THEOREM 9.4 (Artstein). *Let c be a capacity associated with a random set. Then the following statements are equivalent:*

(i) *There is a coupling $(X, \mathcal{G})$ such that $X \sim P, \mathcal{G} \sim c$, and*

$$X \in \mathcal{G} \text{ holds almost surely.} \tag{9.1}$$

(ii) *For every measurable set $B \subseteq \mathcal{X}$,*

$$P(B) \leq c(B). \tag{9.2}$$

This theorem is an important result, which has various implications in probability and statistics, and which is connected to several important concepts. In particular, the set of probability measures P that satisfy inequalities (9.2) for every measurable $B \subseteq \mathcal{X}$ is called the *core* of capacity c.

On a conceptual level, theorem 9.4 provides an answer to problem 9.2 by implying that $\theta \in \Theta_I$ if and only if $P(B) \leq c_{G_\theta(Y)}(B)$ for every measurable set $B \subseteq \mathcal{X}$. However, it does not provide a practical method for determining whether $\theta \in \Theta_I$. Indeed, even with finite $\mathcal{X}$, there are $2^{|\mathcal{X}|}$ inequalities (9.2) to be checked, which is prohibitive.

Instead, the following observation, due to Galichon and Henry [63, 64], that problem 9.2 can be reformulated as a Monge–Kantorovich problem, allows for efficient algorithms to check whether θ is included in the identified set.

THEOREM 9.5. *Consider* $\Phi(x, y) = -1\{x \notin G_\theta(y)\}$, *and define*

$$V(\theta) := \max_{\pi \in \mathcal{M}(P, Q_\theta)} \mathbb{E}_\pi[\Phi(X, Y)]. \tag{9.3}$$

Then statements theorem 9.4(i),(ii) are in turn equivalent to

(iii) the value of $V(\theta)$ of the Monge–Kantorovich problem (9.3) is 0.

As a consequence, the identified set is obtained by

$$\Theta_I = \{\theta \in \Theta : V(\theta) = 0\}.$$

We refer to Galichon and Henry [63, 64] for a proof of this result. Let us show how theorem 9.5(iii) implies Artstein's theorem, that is, theorem 9.4(i),(ii). Indeed, assume $V(\theta) = 0$. Then by Monge–Kantorovich duality, the value of the following program is zero:

$$\inf_{u,v} \mathbb{E}_P[u(X)] + \mathbb{E}_{Q_\theta}[v(Y)] \tag{9.4}$$
$$\text{s.t. } u(x) + v(y) \geq -1\{x \notin G_\theta(y)\}.$$

Following the argument in [148, section 1.4], one can show that one can restrict the maximization problem to $-u$ and v valued in $\{0, 1\}$, that is, $u(x) = -1_B(x)$ and $v(y) = 1_D(y)$. Then, the constraint in problem (9.4) can be reexpressed as

$$1_D(y) \geq 1_B(x) - 1\{x \notin G_\theta(y)\},$$

thus $y \in D^c$ implies $y \in G_\theta^{-1}(B)$. Hence, $D^c \subseteq G_\theta^{-1}(B)$; but as problem (9.4) involves a minimization, one should set D as small as possible, and thus $D = \left(G_\theta^{-1}(B)\right)^c$. Then $v(y) = 1_D(y) = 1\{G_\theta(y) \cap B \neq \emptyset\}$, and $\mathbb{E}_{Q_\theta}[v(Y)] = c_\theta(B)$, where $c_\theta(B) = Q_\theta(G_\theta(Y) \cap B \neq \emptyset)$ is the capacity associated with the random set $G_\theta(Y)$. Hence, the value of the problem is reformulated as

$$V(\theta) = \inf_{B \subseteq \mathcal{X}} c_\theta(B) - P(B),$$

and $V(\theta) = 0$ is equivalent to $P(B) \leq c_\theta(B)$ for all $B \subseteq \mathcal{X}$.

9.1.3 Computation

The computational benefits induced by this formulation as an optimal transport problem are significant. When both $\mathcal{X}$ and $\mathcal{Y}$ are discrete, the value of $V(\theta)$ can be obtained by linear programming. Further, when P and Q_θ put equal weight on every point of their respective supports, combinatorial maximal matching algorithms may be used which in some cases may be more efficient than linear programming algorithms. In applied problems, P is usually an empirical distribution, thus $\mathcal{X}$ is usually a finite sample set. In some cases, Q_θ is discrete; when it is not, one can discretize it as suggested in section 6.5 in order to solve problem 9.2 via these efficient techniques. See [64] for a discussion on various computational methods.

PROGRAMMING EXAMPLE 9.1. The value of $V(\theta)$ given by (9.3) is computed in the program `PerfectMatching.R`. Note that while we have used linear programming techniques (and Gurobi) for convenience, more efficient algorithms exist for these specific types of problems. They are called *maximum cardinality matching algorithms*, and an example is the Hopcroft–Karp algorithm.

9.1.4 References and Notes

The main ideas in this section first appeared in a 2006 manuscript by Galichon and Henry [63] which remained unpublished; two subsequent published references are Beresteanu, Molchanov, and Molinari [15] and Galichon and Henry [64]. Two great references for random sets are the classical book by Matheron [98] and the more recent one by Molchanov [108]. Artstein's theorem appeared in [6]. The reference for Strassen's theorem is [141]; see also [148, section 1.4].

9.2 IDENTIFICATION OF DISCRETE CHOICE MODELS

9.2.1 The Problem

We consider the classical discrete choice model, where heterogenous agents choose some alternative $y \in \mathcal{J} = \{1, \ldots, M\}$. Agents have unobserved heterogeneity in tastes classically represented by some random vector of $\mathbb{R}^M$, $\varepsilon_j \sim P$, where ε_j is interpreted as the stochastic part of the agent's utility if an alternative j is chosen. There is a vector of systematic utilities $(w_j)_{1 \leq j \leq M}$ such that the total utility of a consumer represented by ε if alternative j is chosen is $w_j + \varepsilon_j$. Thus, this consumer chooses alternative j that solves

$$\max_{1 \leq j \leq M} \left\{ w_j + \varepsilon_j \right\}. \tag{9.5}$$

Let q_j be the market share of alternative j, which is the probability that utility from outcome j is higher than the utilities from all the other outcomes. The problem of identification of discrete choice models is formulated as follows.

PROBLEM 9.6. *Assume the vector of choice probabilities (q_j) is observed. What is the set of vectors of systematic utilities (w_j) that are compatible with the discrete choice problem (9.5)?*

We will show that this problem is an instance of the Monge–Kantorovich problem.

9.2.2 The Mass Transport Approach

Let us introduce some notation which will emphasize the connection. Let y_j be the jth vector of the canonical basis of $\mathbb{R}^M$, whose entries are all equal to 0 apart from the jth entry, so that

$$\varepsilon_j = \varepsilon' y_j.$$

Let $\mathcal{Y} = \{y_1, \ldots, y_M\}$. We will extend the notation so that w_y refers to the value of w_j for j so that $y_j = y$; similarly, we will refer indifferently to q_y and q_j. The consumer's indirect utility is

$$\max_{y \in \mathcal{Y}} \left\{ w_y + \varepsilon' y \right\},$$

whose ex ante expected value is given by

$$W(w) = \mathbb{E} \left[\max_{y \in \mathcal{Y}} \left\{ w_y + \varepsilon' y \right\} \right].$$

DEFINITION 9.7. *A vector of systematic utilities w is said to* rationalize *a vector of probabilities (q_y) if q is the probability distribution of some random vector $\tilde{y}$ valued in $\mathcal{Y}$ such that choosing $\tilde{y}$ is optimal for ε, that is,*

$$\tilde{y} \in \arg\max_{y \in \mathcal{Y}} \left\{ w_y + \varepsilon' y \right\}.$$

Equivalently, one says that probability vector q is rationalized by *utility vector w.*

Problem (9.6) therefore boils down to determining the set of vectors of utilities w that rationalize q. As noted in [40], the set of vectors of systematic utilities v that rationalize a given vector of choice probabilities q is the set of v such that $q \in \partial W(w)$, where $\partial W(w)$ is the subdifferential of W at w. This set is conveniently expressed using the convex conjugate of W, as shown in the following result obtained by Chiong, Galichon, and Shum [40].

PROPOSITION 9.8. *The set of utility vectors rationalizing q is $\partial W^*(q)$, which is the subdifferential of W^*, the Legendre–Fenchel transform of W, at q.*

PROOF. It follows from the envelope theorem that q is rationalized by w if and only if $q \in \partial W(w)$. By the equivalence between (6.8) and (6.9) spelled out in section 6.1, this is equivalent to $w \in \partial W^*(q)$. □

It turns out that $W^*(q)$ and its subdifferential $\partial W^*(q)$ are very closely connected to the theory of optimal transport. Indeed, the following result from Galichon and Salanié [69] shows that $-W^*(q)$ is the value of the Monge–Kantorovich problem, while $\partial W^*(q)$ is the opposite of the set of potentials (w_y) that appear in the solution to the dual formulation of that problem.

THEOREM 9.9. *The set of vectors of systematic utilities (w_y) that rationalize a given vector of choice probabilities q is the set of vectors $(-v_y)$, where v are the minimizers of the problem*

$$\min_{v \in \mathbb{R}^M} \mathbb{E}\left[\max_{y \in \mathcal{Y}} \{-v_y + \varepsilon' y\}\right] + \sum_{y \in \mathcal{Y}} q_y v_y. \tag{9.6}$$

Further, the value of problem (9.6) coincides with $-W^*(q)$.

PROOF. By definition of the convex conjugate,

$$W^*(q) = \max_w \left\{ \sum_{y \in \mathcal{Y}} q_y w_y - \mathbb{E}\left[\max_{y \in \mathcal{Y}} \{w_y + \varepsilon' y\}\right] \right\};$$

hence,

$$W^*(q) = -\min_w \left\{ -\sum_{y \in \mathcal{Y}} q_y w_y + \mathbb{E}\left[\max_{y \in \mathcal{Y}} \{w_y + \varepsilon' y\}\right] \right\}$$

$$= -\min_v \left\{ \sum_{y \in \mathcal{Y}} q_y v_y + \mathbb{E}\left[\max_{y \in \mathcal{Y}} \{-v_y + \varepsilon' y\}\right] \right\}.$$

□

Hence, the problem of identification of a discrete choice system is precisely the same problem as the one described in theorem 5.2, which is an instance of the Monge–Kantorovich problem.

9.2.3 Computation

The previous discussion suggests a method for identifying the vector of systematic utilities w (or equivalently, the vector v) when the distribution of ε is continuous: sample the distribution P at N sample points, and solve the optimal assignment problem which is the sample analogue of (9.6). This approach, initiated by [40], has been called the *mass transport approach* to identification of discrete choice models. The following result, obtained by Chiong, Galichon, and Shum [40], ensures consistency of this technique as $N \to +\infty$.

THEOREM 9.10. *Consider* $\{\varepsilon^1, \ldots, \varepsilon^N\}$, *an i.i.d. sample from P, which is assumed to be continuous and to have full support. Let* (u_i^N, v_j^N) *be a solution to the dual optimal assignment problem*

$$\min_{u,v} \frac{1}{N} \sum_{i=1}^{N} u_i + \sum_{j=1}^{M} q_j v_j$$

$$\text{s.t. } u_i + v_j \geq \varepsilon_j^i \quad \forall i \in \{1, \ldots, N\}, \forall j \in \{1, \ldots, M\}.$$

Then $N \to +\infty$, *and (up to an additive constant)* v_j^N *converges almost surely to the solution to (9.6), which is unique (again, up to an additive constant).*

PROOF. See [40, theorem 5]. □

PROGRAMMING EXAMPLE 9.2. The R package TraME [70], contains a number of resources for identification and equilibrium computation of discrete choice and matching models, based on the optimal transportation methods discussed in these notes.

9.2.4 References and Notes

The references for this section are Galichon and Salanié [69] and Chiong, Galichon, and Shum [40]. The reference for TraME is [70].

9.3 HEDONIC EQUILIBRIUM

9.3.1 The Problem

We consider a discrete version of the hedonic equilibrium problem. Consider the market for wine. The producers and the consumers are characterized by their type, respectively $x \in \mathcal{X}$ and $y \in \mathcal{Y}$. The wine may come in different qualities $z \in \mathcal{Q}$. It is assumed that the total production of producers of type x is p_x and the total consumption of consumers of type y is q_y. Total production equals total consumption, thus

$$\sum_{x \in \mathcal{X}} p_x = \sum_{y \in \mathcal{Y}} q_y.$$

Let w_z be the price of a unit of wine of quality z (which is an endogenous quantity determined at equilibrium). It is assumed that the surplus of a producer x per unit z sold is $\alpha_{xz} + w_z$, where α is the producer's ability, which is essentially the opposite of the production costs. This term depends on both x and z: some producers are better than others in different wines. The producer's indirect surplus is thus

$$u_x = \max_{z \in \mathcal{Q}} \{\alpha_{xz} + w_z\}. \tag{9.7}$$

Likewise, the surplus of a consumer y per unit z consumed is $\gamma_{zy} - w_z$, which is quasi-linear in money; γ_{zy} is essentially the monetary valuation of wine of quality z by a consumer of type y. Therefore, the consumer's indirect utility is

$$v_y = \max_{z \in \mathcal{Q}} \{\gamma_{zy} - w_z\}. \tag{9.8}$$

Let π_{xz} be the quantity of units of type z produced by producers of type x, and let π_{zy} be the quantity of units of type z consumed by consumers of type y. One has

$$\sum_{z \in \mathcal{Q}} \pi_{xz} = p_x \quad \text{and} \quad \sum_{z \in \mathcal{Q}} \pi_{zy} = q_y, \tag{9.9}$$

and the optimal surplus is given by

$$\max_{\pi \geq 0} \sum_{x \in \mathcal{X}, z \in \mathcal{Q}} \pi_{xz} \alpha_{xz} + \sum_{z \in \mathcal{Q}, y \in \mathcal{Y}} \pi_{zy} \gamma_{zy} \tag{9.10}$$

$$\text{s.t. (9.9)}.$$

DEFINITION 9.11 (Hedonic equilibrium). *A hedonic equilibrium outcome is the specification of* $(\pi_{xz}, \pi_{zy}, w_z)$, *where* π_{xy} *is the breakdown of the production,* π_{zy} *is the breakdown of the consumption, and* (w_z) *is the equilibrium price of each of the qualities, such that*

(i) *π solves (9.9);*
(ii) *the indirect surpluses (u_x) and (v_y) defined by (9.7) and (9.8) are such that $\pi_{xz} > 0$ implies $u_x = \alpha_{xz} + w_z$ and $\pi_{zy} > 0$ implies $v_y = \gamma_{zy} - w_z$.*

The problem we will address is the existence and computation of an equilibrium in this setting.

PROBLEM 9.12. *Does a hedonic equilibrium exist? If so, how can it be computed?*

9.3.2 Network Flow Reformulation

Maurice Queyranne proposed the following reformulation as a network flow problem. Consider $(\mathcal{Z}, \mathcal{A}, n, \Phi)$, a network such that

$$\mathcal{Z} = \mathcal{X} \cup \mathcal{Y} \cup \mathcal{Q},$$

$$\mathcal{A} = \mathcal{X} \times \mathcal{Q} \cup \mathcal{Q} \times \mathcal{Y},$$

$$n_z = q_z 1\{z \in \mathcal{Y}\} - p_z 1\{z \in \mathcal{X}\},$$

$$\Phi_{zz'} = \alpha_{zz'} 1\{zz' \in \mathcal{X} \times \mathcal{Q}\} + \gamma_{zz'} 1\{zz' \in \mathcal{Q} \times \mathcal{Y}\}.$$

This network is regular, and the optimal flow problem is given by

$$\max_{\pi \geq 0} \sum_{a \in \mathcal{A}} \pi_a \Phi_a \tag{9.11}$$

$$\text{s.t. } \nabla^* \pi = n,$$

whose dual is

$$\min_{\tilde{v}} \sum_{z \in \mathcal{Z}} n_z \tilde{v}_z \tag{9.12}$$

$$\text{s.t. } \nabla \tilde{v} \geq \Phi.$$

The following result argues that the solution to this problem provides the hedonic equilibrium outcome.

THEOREM 9.13 (Queyranne). *Let π and $\tilde{v}$ be solutions to the primal (9.11) and the dual (9.12), respectively. Let $u_x = -\tilde{v}_x$ for $x \in \mathcal{X}$, $w_z = -\tilde{v}_z$ for $z \in Q$, and $v_y = \tilde{v}_y$ for $y \in \mathcal{Y}$. Then $\left(\pi_{xz}, \pi_{zy}, w_z\right)$ is a hedonic equilibrium outcome, and (u_x) and (v_y) are the associated indirect surpluses of producers and consumers, respectively.*

As a consequence, producers and consumers match according to a matching surplus $\Phi(x, y)$ equal to the optimal path between producer x and consumer y, that is,

$$\Phi(x, y) = \max_{z \in Q} \left\{ \alpha_{xz} + \gamma_{yz} \right\}. \tag{9.13}$$

9.3.3 Computation

We will use the sparse structure of the network. Then Φ_a will result from the concatenation of $\text{vec}(\alpha)$ and $\text{vec}(\gamma)$.

PROGRAMMING EXAMPLE 9.3. The program HedonicEquilibrium.R implements the computation of a hedonic equilibrium by linear programming, using techniques similar to those employed by programming example 8.1.

9.3.4 References and Notes

Standard references for the modern theory of hedonic models are Ekeland, Heckman, and Nesheim [52], who build on classical work by Tinbergen [143] and Rosen [124]. See also Ekeland's lecture notes [49]. Theorem 9.13 is an unpublished idea from Queyranne [116]. Formula (9.13) linking hedonic and matching models was discovered independently by Chiappori, McCann, and Nesheim [39]. An entropic regularization of the network flow problem (9.11) was provided in [48] for identification purposes.

9.4 IDENTIFICATION VIA VECTOR QUANTILE METHODS

9.4.1 The Problem

Consider a continuous hedonic model similar to the one discussed in the previous section, but with continuous characteristics. We have a consumer, of observed characteristics $x \in \mathbb{R}^k$ and latent vector of characteristics $u \in \mathbb{R}^d$, choosing a good (say, a house) characterized by a vector of attributes $y \in \mathbb{R}^d$ (say, size and various amenities). Assume the utility of the consumer choosing y is given by $S(x, y) + u'y$, where $S(x, y)$ is the observed part of the consumer surplus, which is assumed to be concave in y, and where $u'y$ is a preference shock. The indirect utility of the consumer is given by

$$\varphi(x, u) = \max_{y} \left\{ S(x, y) + u'y \right\}, \tag{9.14}$$

so by first-order conditions, $\partial S(x, y)/\partial y_i + u_i = 0$ for each $1 \leq i \leq d$; thus, letting $\psi(x, y) = -S(x, y)$, quality y is chosen by consumer (x, u) such that $u_i = \partial \psi(x, y)/\partial y_i$. The problem is the following.

PROBLEM 9.14. *Assume that u is distributed independently from x with a known distribution μ (say, $\mathcal{U}([0,1])$). We observe the distribution of choices Y given characteristics X = x. Can one recover the vector of marginal surpluses $\partial S(x, y)/\partial y_i$?*

9.4.2 Conditional Vector Quantiles

In the case of scalar characteristics ($d = 1$), pioneering work by Rosa Matzkin [99] showed the relevance of *quantile methods* to addressing this question. In this case, for each x, $\partial \psi(x, y)/\partial y$ is the unique nondecreasing map pushing forward the distribution of $Y|X = x$ on the distribution of $U|X = x$, which is the uniform distribution on $[0, 1]$. Thus ψ is identified up to an additive function of x by the following theorem.

THEOREM 9.15 (Matzkin). *One has $\partial\psi/\partial y(x, y) = F_{Y|X}(y|x)$, hence the marginal surplus $\partial_y S(x, y)$ is identified. By the same token, the indirect surplus $\varphi(x, u)$ is identified up to an additive function of x by $\partial\varphi/\partial u(x, u) = F_{Y|X}^{-1}(u|x)$.*

Our goal here is to show how Matzkin's identification strategy via the quantile approach can be generalized to the multivariate case ($d \geq 1$). We shall demonstrate the usefulness of a notion of vector quantile based on Brenier–McCann's theorem. By first-order conditions in (9.14), the choice of quality vector y by a consumer of type x is rationalized by a latent vector $u(x, y)$ such that

$$u(x, y) := \nabla_y \psi(x, y),$$

which, conditional on x, is "vector nondecreasing" in y, by which we mean the "gradient of a convex function." Omitting the dependence in x, the problem consists of finding out whether for $Y \sim F_Y$, there is a convex function $\psi(y)$ such that

$$\nabla\psi(Y) \sim \mu.$$

The answer is yes, by Brenier–McCann's theorem (theorem 6.7). When μ and F_Y have finite second moments, ψ is the solution to

$$\min_{\psi,\varphi} \mathbb{E}_\mu[\varphi(U)] + \mathbb{E}_{F_Y}[\psi(Y)] \tag{9.15}$$
$$\text{s.t. } \varphi(u) + \psi(y) \geq u'y,$$

which is the dual Monge–Kantorovich problem whose primal formulation is

$$\max_\pi \mathbb{E}_\pi[U'Y], \tag{9.16}$$
$$\pi \in \mathcal{M}(\mu, F_Y),$$

and by theorem 6.5, both (9.15) and (9.16) have solutions, which are convex and are related by $U = \nabla\psi(Y)$ and $Y = \nabla\varphi(U)$. Note that $\nabla\psi$ is the F_Y-vector quantile of μ, and $\nabla\varphi$ is the μ-vector quantile of F_Y in the sense of definition 6.8.

Now let us go back to the conditional case. Integrating problem (9.15) over x, we get

$$\min_{\psi,\varphi} \mathbb{E}_{F_{XY}}[\psi(X, Y)] + \mathbb{E}_{F_X \otimes \mu}[\varphi(X, U)] \tag{9.17}$$
$$\text{s.t. } \psi(x, y) + \varphi(x, u) \geq u'y,$$

whose solutions $\varphi(x, \cdot)$ and $\psi(x, \cdot)$ are conjugate in the sense that

$$\varphi(x, u) = \sup_y \{-\psi(x, y) + u'y\},$$

$$\psi(x, y) = \sup_u \{-\varphi(x, u) + u'y\}. \tag{9.18}$$

Problem (9.17) has a primal formulation which is

$$\max_{(X,Y,U)\sim\pi} \mathbb{E}_\pi [U'Y]$$

$$\text{s.t. } (X, Y) \sim F_{XY}, \tag{9.19}$$

$$U \sim \mu,$$

$$U, X \text{ independent.}$$

The solutions to problems (9.15) and (9.16) are now related by $U = \nabla_y\psi(X, Y)$ and $Y = \nabla_u\varphi(X, U)$. This brings the full answer to problem 9.14 above. Carlier, Chernozhukov, and Galichon [33] define $\nabla_u\varphi(x, \cdot)$ as the μ-conditional vector quantile of Y given x, and $\nabla_y\psi(x, \cdot)$ as the μ-conditional inverse vector quantile of Y given x. This led Chernozhukov et al. [38] to formulate the following result.

THEOREM 9.16. *The vector of marginal indirect surpluses is identified as the μ-conditional vector quantile of F_Y given x, and the vector of marginal surpluses $\nabla_y S(x, y)$ is identified as the opposite of the μ-conditional inverse vector quantile of Y given x. Both of these quantities are computed using problem (9.17).*

Computational issues are discussed in section 9.5.

9.4.3 References and Notes

Quantile methods for identification were pioneered by Matzkin [99], and applied to identification in hedonic models by Heckman, Matzkin, and Nesheim [78]. See a survey of related methods in [100]. These ideas were carried onward to the case with multivariate attributes by Chernozhukov et al. [38].

9.5 VECTOR QUANTILE REGRESSION

9.5.1 The Problem

We saw in section 9.4 that the marginal willingness to pay in problem 9.14, as well as the marginal indirect surplus, are identified respectively by the F_Y- *conditional vector quantile* of μ given x, and $\nabla_u\varphi(x, \cdot)$ as the μ-conditional vector quantile of F_Y given x. While these quantities are theoretically obtainable by problem (9.17), estimation will soon encounter practical difficulties, once x is continuously distributed. Indeed, to solve problem (9.17) in practice, one relies on a discretization of U on a grid with M points, leading to a sample $\{U_1, \ldots, U_M\}$, and one uses the sample distribution of (X, Y), namely, $\{(X_1, Y_1), \ldots, (X_N, Y_N)\}$. However, keep in mind that problem (9.17) is simply the integration of the conditional version of problem (9.15) with respect to X; but, if the distribution of X is continuous, then each value taken by X in the sample is unique, and thus the sample version of the conditional problem (9.15) is uninformative. To get away from this problem, one way out is to consider a parametric version of problem (9.15), where the dependence of φ on x is assumed to belong to a parametric family. The problem addressed in the present section is therefore the following one.

PROBLEM 9.17. *How can the definition and the computation of conditional vector quantiles be adapted so that the dependence on the conditioning variable is parametric?*

9.5.2 The Linear Specification

Consistent with quantile regression, we will make the assumption that φ is linear in x, that is,

$$\varphi(u, x) = x' b(u). \tag{9.20}$$

This assumption is without much loss of generality, as one can always expand x by adding nonlinear functions of the entries of x. More importantly, we will assume that the first component of X is 1, which will mean that an intercept term is included in the regression model. The dimension of X is denoted by p and we shall denote $X = (1, X_{-1})$ by $X_{-1} \in \mathbb{R}^{p-1}$. We set $\bar{x} = E[X]$. Recall that the μ-conditional vector quantile of Y given $X = x$ is $\nabla_u \varphi(u, x)$. Hence imposing (9.20) yields

$$y = Db(u)' x, \tag{9.21}$$

where, as we recall, Db is the Jacobian matrix of b. In dimension $d = 1$ and when $\mu = \mathcal{U}([0, 1])$, this expression coincides with the usual specification of linear quantile regression $y = x' \beta(u)$, where $\beta(u)$ is simply the derivative of b at u.

If (9.20) is imposed in problem (9.15), the latter becomes

$$\min_{\psi, b} \mathbb{E}_{F_{XY}} [\psi(X, Y)] + \bar{x}' \mathbb{E}_\mu [b(U)]$$

$$\text{s.t. } \psi(x, y) + x' b(u) \geq u' y, \tag{9.22}$$

and as before, we may express ψ as a function of b and get

$$\psi(x, y) = \sup_{y} \left\{ u'y - x'b(u) \right\},$$

whose first-order conditions are (9.21). As before, problem (9.22) has a dual formulation. The corresponding primal formulation is

$$\max_{(X,Y,U)\sim\pi} \mathbb{E}_{\pi}\left[U'Y \right]$$

$$\text{s.t. } (X, Y) \sim F_{XY},$$

$$U \sim \mu,$$

$$\mathbb{E}_{\pi}\left[X|U \right] = \bar{x}.$$

(9.23)

Of course, the problem can be equivalently stated in terms of L^2- minimization $\min \mathbb{E}_{\pi}\left[|U - Y|^2 \right]$ under the same set of constraints. Note that problem (9.23) is closely related to problem (9.19), but in contrast to that problem, the constraint of independence between X and U is replaced by a constraint of mean-independence of X from U.

We now make formal assumptions under which the linear vector quantile regression model is specified.

ASSUMPTION 9.18. *Assume that*

(a) *μ and F_Y have finite second moments;*
(b) *μ is continuous and supported on a convex set $\mathcal{U} \subseteq \mathbb{R}^d$;*
(c) *for each x in the support, $F_{Y|X=x}$ is continuous;*
(d) *we have a quasi-linear representation a.s.,*

$$Y = \beta\left(\tilde{U}\right)' X, \quad \tilde{U} \sim \mu, \quad \mathbb{E}\left[X|\tilde{U}\right] = \bar{x},$$

(9.24)

where $u \mapsto \beta(u)$ is a map from $\mathcal{U}$ to the set of $p \times d$ matrices such that $u \mapsto \beta(u)'x$ is a gradient of a convex function for each x in the support, and a.e. $u \in \mathcal{U}$,

$$x'\beta(u) = \nabla_u \Phi_x(u), \quad \Phi_x(u) := x'B(u),$$

where $u \mapsto B(u)$ is a C^1-map from $\mathcal{U}$ to $\mathbb{R}^d$, and $u \mapsto B(u)'x$ is a strictly convex map from $\mathcal{U}$ to $\mathbb{R}$.

Then Carlier, Chernozhukov, and Galichon [33] have shown the following result.

THEOREM 9.19. *Under assumption 9.18, the following holds:*

(i) *The random vector $\tilde{U}$ entering the quasi-linear representation (9.24) solves (9.23).*

(ii) *The quasi-linear representation is unique a.s., that is, if we also have*
$Y = \bar{\beta}(\bar{U})'X$ *with* $\bar{U} \sim F_U$, $\mathbb{E}\left[X|\bar{U}\right] = \bar{x}$, *and* $u \mapsto \bar{\beta}(u)'X$ *is a gradient*
of a strictly convex function in $u \in \mathcal{U}$ *a.s., then* $\bar{U} = \tilde{U}$ *and* $\beta(\tilde{U})'X =$
$\bar{\beta}(\tilde{U})'X$ *a.s.*

9.5.3 Link with Classical Quantile Regression

When $d = 1$, it is possible to show that problem (9.23) is equivalent to the
problem

$$\max_{A_t \geq 0} \int_0^1 \mathbb{E}[A_u Y]du \tag{9.25}$$

$$\text{s.t. } A_t \leq 1\,[P]\,, \tag{9.26}$$

$$\mathbb{E}[A_t X] = (1 - t)\,\bar{x}\,[\beta_t]\,, \tag{9.27}$$

$$A_t \leq A_s,\ t \geq s, \tag{9.28}$$

where the primal variables are related by $U = \int_0^1 A_\tau\,d\tau$, and the dual variables
by $\psi(x, y) = \int_0^1 \left(y - x'\beta_\tau\right)^+ d\tau$ and $b(t) = \int_0^t \beta_\tau d\tau$. This connection is very
interesting. Indeed, problem (9.25) without the final constraint (9.28) is exactly
classical quantile regression. Vector quantile regression consists therefore of
adding the constraint that $t \to A_t$ is nonincreasing. However, in classical
quantile regression, $A_t = 1\{Y \geq X'\beta_t^{QR}\}$, thus this constraint is automatically satisfied if $t \to x'\beta_t^{QR}$ is nondecreasing for every x in the support, that
is, if classical quantile regression does not exhibit the "crossing problem."

9.5.4 Computation

In practical applications (X, Y) is usually given as an empirical sample of size
n. We will discretize μ into m sample points. Let p be the number of regressors.
The LP formulation of the discretized problem is thus

$$\max_{\pi \geq 0} \text{Tr}(U'\pi Y)$$

$$\text{s.t. } 1'_m \pi = \nu'\,[\psi']\,,$$

$$\pi X = \mu\bar{x}\,[b]\,,$$

where X is an $n \times p$ matrix, Y is $n \times d$, ν is $n \times 1$ such that ν_i is the weight of
observation (X_i, Y_i), U is $m \times d$, μ is $m \times 1$, and π is $m \times n$. Vectorizing the

objective function and the constraints yields

$$\mathrm{Tr}\left(U'\pi Y\right) = \mathrm{vec}\left(UY'\right)' \mathrm{vec}(\pi),$$
$$\mathrm{vec}\left(1'_m\pi\right) = \left(I_n \otimes 1'_m\right) \mathrm{vec}(\pi),$$
$$\mathrm{vec}\left(\pi X\right) = \left(X' \otimes I_m\right) \mathrm{vec}(\pi).$$

PROGRAMMING EXAMPLE 9.4. This approach is implemented in the program VQR.R which implements the previous discretized approach using Gurobi. The files test_VQR_Engel1D.R and test_GDPDebtDeficit.R respectively provide examples with $d = 1$ and $d = 2$.

9.5.5 References and Notes

A great reference for classical quantile regression is the 2005 textbook [88] by Koenker, who is the father of quantile regression. The reference for vector quantile regression is Carlier, Chernozhukov, and Galichon [33].

9.6 IMPLEMENTABLE MECHANISMS

9.6.1 The Problem

Consider the problem of a mobile phone operator (the principal) selling a service (access to the cellular network) to a client (the agent). The agent's type is $x \in \mathcal{X} \subseteq \mathbb{R}^d$, and the service may come in various qualities $y \in \mathcal{Y} \subseteq \mathbb{R}^d$, which is interpreted as the specifics of the plan: number of hours of airtime included, volume of data, international options, lock-in period, etc. Assume that the cost for the principal of producing service y is $c(y)$. An agent of type x has valuation $\Phi(x, y)$ for a service of type of quality y, and has reservation utility $u_0(x)$ if the service is not purchased. It is assumed that $\Phi(x, y) \geq u_0(x)$ for every x and y.

9.6.1.1 First-Degree Price Discrimination

If personalized pricing, or *first-degree price discrimination*, is allowed, and if the principal is aware of the agent's type, then the price charged by the principal for a service of type y to an agent of type x is designed to extract the whole surplus, and make the agent indifferent between getting the service and opting out; hence, that personalized price is

$$V(x, y) = \Phi(x, y) - u_0(x).$$

As agents are indifferent between any service offered (which all provide them with zero surplus), the principal can use infinitesimal price variation to induce each agent to choose the profit maximizing quality, so that the profit extracted by the principal from an agent of type x is

$$S(x) = \max_{y \in \mathcal{Y}} \{\Phi(x, y) - c(y)\} - u_0(x),$$

and, assuming the agents' types are distributed according to P, the overall profit of the monopolist under first-degree price discrimination is

$$\mathbb{E}_P \left[\max_{y \in \mathcal{Y}} \{\Phi(X, y) - c(y)\} \right] - \mathbb{E}_P [u_0(X)].$$

Thus, if the principal is in a monopolistic situation and has perfect information on agents' types, and if first-degree price discrimination is allowed, the agents will be left with zero surplus. This very undesirable situation is limited by two facts. The first one is that many countries explicitly forbid first-degree price discrimination, especially in monopolistic or oligopolistic markets. The second one is that the principal often has limited or no information on agents' types.

9.6.1.2 Second-Degree Price Discrimination

When first-degree price discrimination is forbidden, *second-degree price discrimination* is usually allowed. This practice consists of offering a menu of services $y \in \mathcal{Y}$ at prices $v(y)$ which are differentiated among the services, but not the consumers. In this case, assuming that the prices are set such that $\Phi(x, y) - v(y) \geq u_0(x)$ for all x and y, agent x's choice problem will be

$$\max_{y \in \mathcal{Y}} \{\Phi(x, y) - v(y)\},$$

and the agent will choose quality $y = T(x)$ so that

$$T(x) \in \arg\max_{y \in \mathcal{Y}} \{\Phi(x, y) - v(y)\}.$$

9.6.1.3 Asymmetric Information

Let us now consider the situation when first-degree price discrimination is allowed, but the principal has no information on the agent's type. Then the agent will announce a type $x' \in \mathcal{X}$, but it will be impossible for the principal to verify it. The recourse for the principal is then to induce an agent announcing type x' to purchase service $T(x')$ at price $v(T(x'))$. This can be done, for instance, by choosing $V(x', y)$ large enough as soon as $y \neq T(x')$. We should stress here that in this interpretation, the allocation rule T and the payment rule v are chosen by the principal. This motivates the following definition.

DEFINITION 9.20. *A mechanism* (T, v) *consists of the specification of an allocation rule* $y = T(x) \in \mathcal{Y}$, *and a* payment $v(y)$ *required from the agent.*

The agent may choose which type to announce. Let $X(x)$ be the type announced by an agent of type x; one has

$$\Phi\left(x, T\left(X(x)\right)\right) - v\left(T\left(X(x)\right)\right) \geq \Phi(x, T(x')) - v(T(x')). \qquad (9.29)$$

Equivalently, the principal may take over the task of converting x into $x' = X(x)$, and may use allocation rule $T \circ X$ instead of T, in which case the agent will announce their true type: this is the *revelation principle*; see [151] and references therein. Thus, without loss of generality, we may assume $X(x) = x$ in (9.29). This leads to the following definition.

DEFINITION 9.21 (Implementability). *Mechanism* (T, v) *is implementable in dominant strategy, or simply implementable, if for all* $(x, x') \in \mathcal{X}^2$,

$$\Phi(x, T(x)) - v(T(x)) \geq \Phi(x, T(x')) - v(T(x')).$$

Allocation rule T *is implementable if and only if there exists a payment rule* v *such that mechanism* (T, v) *is implementable.*

Given an allocation rule T, it is far from obvious that T is implementable. Therefore, we have our first question.

PROBLEM 9.22. *Given an allocation rule* T, *how can we check whether* T *is implementable, and, if it is, can we determine the set of payment rules* v *such that mechanism* (T, v) *is implementable?*

Our next question is the choice of the optimal mechanism for the monopolist.

PROBLEM 9.23. *Assume that first-degree price discrimination is illegal, or that it is legal but that the principal has no information about agents' types (although it knows its distribution). What is the optimal implementable mechanism from the point of view of the principal?*

Problem 9.22 will be addressed in section 9.6.2, while problem 9.23 will be dealt with in section 9.6.3.

9.6.2 Implementability

We can always redefine $\mathcal{Y}$ as $\mathcal{Y} = T(\mathcal{X})$. Doing so, we see that allocation rule T is implementable if and only if there exists a payment rule $v(\cdot)$ such that

$$T(x) \in \arg\max_{y \in \mathcal{Y}} \{\Phi(x, y) - v(y)\}.$$

Let us now relate this to optimal transport and generalized convexity. Consider a probability measure P on $\mathcal{X}$ (which is the measure over agents' types), and let Q be the measure over the set of outcomes $\mathcal{Y}$ such that $Q = T\#P$. Hence, Q is the measure over the outcomes that are actually decided. Our main result for this section is a general characterization of implementable maps using the theory of optimal transport. Let P be a continuous distribution on $\mathcal{X}$, and let Q be the image measure of P by T, that is, the distribution of $T(X)$, where $X \sim P$.

THEOREM 9.24 (Carlier). *The following statements are equivalent:*

(i) *The map $T : \mathcal{X} \to \mathcal{Y}$ is implementable.*
(ii) *The map T solves the Monge problem associated with Φ between measures P and $Q = T\#P$, that is,*

$$\mathbb{E}_P\left[\Phi\left(X, T(X)\right)\right] = \max_{\pi \in \mathcal{M}(P,Q)} \mathbb{E}_\pi\left[\Phi(X, Y)\right]. \qquad (9.30)$$

Further, when this is the case, the payment rules such that (T, v) is implementable are such that there is some function u such that (u, v) is a solution to the dual problem

$$\min_{u,v} \mathbb{E}_P\left[u(X)\right] + \mathbb{E}_P\left[v\left(T(X)\right)\right]$$
$$\text{s.t. } u(x) + v(y) \geq \Phi(x, y). \qquad (9.31)$$

From this equivalence we see that testing whether a given map T is implementable is equivalent to an optimal transportation problem. When Φ satisfies the conditions in chapter 7, then for P-almost every x,

$$\nabla u(x) = \frac{\partial \Phi\left(x, T(x)\right)}{\partial x}, \qquad (9.32)$$

where u appears in the solution to (9.31). This suggests that the payment v is determined from the knowledge of T. Indeed, condition (9.32) implies that u is determined up to a constant; next, v is determined on $\mathcal{Y} = T(\mathcal{X})$ by

$$v\left(T(x)\right) = \Phi\left(x, T(x)\right) - u(x)$$

(this determination is still up to a constant).

Note that in the one-dimensional case and when Φ satisfies the Spence–Mirrlees condition $\partial^2 \Phi / \partial x\, \partial y > 0$, it is well known that T is implementable if and only if T is nondecreasing. Nondecreasing functions are known to be optimal transport maps in the problem with supermodular surplus.

In particular, when $\Phi(x, y) = x'y$, then the implementable functions are the gradients of convex functions, recovering a result from Rochet [119] and McAfee and McMillan [101]. In particular, when the space of outcomes is a finite set Ω, the set $\mathcal{Y}$ may be taken as the canonical basis of $\mathbb{R}^{\Omega}$, and x_{ω} is the agents' valuation of outcome ω. In this case, if $y = e^{\omega}$ is the element of the canonical basis associated with ω, then $x'y = x'e^{\omega} = x_{\omega}$. Then the implementable rules are of the form

$$T(x) \in \arg\max_{y \in \mathcal{Y}} \left\{ x'y - v(y) \right\};$$

that is, they are of the form $T(x) = \nabla u(x)$, where $u(x) = v^*(x)$. Hence, the set of implementable allocations comprises the gradients of piecewise affine convex functions.

9.6.3 The Monopolist Problem

Consider the case of a monopolist who produces a variety of services $y \in \mathcal{Y}$ at cost $c(y)$. The monopolist offers the service at price $v(y)$, and the corresponding profit is thus $v(y) - c(y)$. The monopolist knows the distribution P of consumers' types in the population. If (T, v) is implementable, then a consumer of type x will choose $y = T(x)$. Further, it is assumed that a consumer of type x will get an outside utility equal to $u_0(x)$, and thus $u(x) \geq u_0(x)$, hence

$$\max_y \{\Phi(x, y) - v(y)\} \geq u_0(x) \quad \forall x \in \mathcal{X}. \tag{9.33}$$

Thus, the overall profit of the monopolist will be $\mathbb{E}_P\left[v\left(T(X)\right) - c\left(T(X)\right)\right]$, and the monopolist's program is therefore

$$\max_{(T,v)} \mathbb{E}_P\left[v\left(T(X)\right) - c\left(T(X)\right)\right]$$

$$\text{s.t. } (T, v) \text{ is implementable,}$$

$$(9.33) \text{ holds.}$$

However, note that $v(T(X)) = \Phi(X, T(X)) - u(X)$; thus the problem becomes

$$\max_u \mathbb{E}_P[\Phi(X, \partial^{\Phi} u(X)) - u(X) - c(\partial^{\Phi} u(X))]$$

$$\text{s.t. } u \text{ is } \Phi\text{-convex,}$$

$$u \geq u_0,$$

where $\partial^{\Phi} u$ is defined in chapter 7 as the Φ-gradient of u. Note that this problem is a delicate variational problem, which is highly nonlinear. The question of whether the set of constraints is convex, which boils down to the question of whether the set of Φ-convex functions is convex, is an interesting and nontrivial question tackled by Figalli, Kim, and McCann in [56].

In the case $\Phi(x, y) = x'y$, which was the case originally considered by Rochet and Choné [120], the problem boils down to

$$\max_{u} \mathbb{E}_P \left[X' \nabla u(X) - u(X) - c\left(\nabla u(X)\right) \right]$$

s.t. u is convex,

$u \geq u_0.$

Even in this particular case, the convexity condition is not easy to handle from a numerical standpoint. Recent progress has been made in this direction; see [107].

9.6.4 References and Notes

A general reference on price discrimination and implementability is Laffont [91]. Relevant references for section 9.6.2 are Rochet [119], McAfee and McMillan [101], and the papers by Carlier [29–31]. These ideas are extended to the non-quasi-linear setting by Noldeke and Samuelson in a recent preprint [113]; see also the discussion in the conclusion.

The main reference for section 9.6.3 is the original paper by Rochet and Choné [120], building on previous work by Mussa and Rosen [111] and Armstrong [4]; see also Armstrong and Rochet [5] and Rochet and Stole [121]. More recently, Figalli, Kim, and McCann [56] have provided geometric conditions on the surplus Φ for the set of Φ-convex functions to be convex; see also [103]. Progress on numerical methods has been made with the work of Mérigot and Oudet [107].

9.7 NO-ARBITRAGE PRICING OF FINANCIAL DERIVATIVES

9.7.1 The Problem

Consider the problem of a bank who wants to price an option of maturity $T = 1$ on two underlyings X and Y. The payoff of the option at date $T = 1$ is

$$\Phi(X, Y);$$

for example, spread options $\Phi(X, Y) = (X - Y - k)^+$, cheapest to deliver $\Phi(X, Y) = \min(X, Y)$, etc. We shall assume there is a perfectly liquid and complete market of single-name vanilla options on X and Y, so that the risk-neutral marginal probabilities P of X and Q of Y are fully known. Let $\mathcal{M}(P, Q)$ be the set of risk-neutral probabilities with these marginals. We state the problem.

PROBLEM 9.25. *What is the set of prices for the option that does not generate an arbitrage opportunity?*

9.7.2 Arbitrage Bound and Superreplicating Portfolio

Without loss of generality, we will focus on the upper bound. The upper arbitrage bound on the option price V is

$$V^{\max} = \max_{\pi \in \mathcal{M}(P,Q)} \mathbb{E}_\pi \left[\Phi(X, Y) \right], \tag{9.34}$$

which is quite obviously a Monge–Kantorovich problem. Without loss of generality, consider the upper bound. Interestingly, the dual Monge–Kantorovich problem has an interpretation in terms of a *superreplicating portfolio*. A classical argument due to Breeden and Litzenberger [23] argues that any contingent profile $u(X)$ can be synthesized as a combination of puts and call options on X. As the only traded assets are puts and call options on X and Y, this implies that the profiles that can be synthesized are the profiles of the type $u(X) + v(Y)$. Because the risk-neutral probability of X is P, the price of the part of the profile yielding $u(X)$ is $\mathbb{E}_P [u(X)]$, while the price of the part of the profile yielding $v(Y)$ is $\mathbb{E}_Q [v(Y)]$. The price of a profile yielding $u(X) + v(Y)$ is therefore $\mathbb{E}_P \lfloor u(X) \rfloor + \mathbb{E}_Q [v(Y)]$.

DEFINITION 9.26. *A superreplicating portfolio of a contingent claim ζ is a portfolio of traded assets whose contingent payoff is always weakly greater than ζ.*

Consider the problem of the bank which has sold the option. Its contingent liability is $\Phi(X, Y)$, which it may wish to *superhedge*, by purchasing a superreplicating portfolio. Given the discussion above, a superreplicating portfolio is a profile $u(X) + v(Y)$ such that

$$u(X) + v(Y) \geq \Phi(X, Y),$$

while its price is $\mathbb{E}_P [u(X)] + \mathbb{E}_Q [v(Y)]$. The price of the cheapest superreplicating portfolio is therefore

$$\min_{u,v} \mathbb{E} [u(X)] + \mathbb{E} [v(Y)]$$

$$\text{s.t. } u(x) + v(y) \geq \Phi(x, y),$$

which is exactly the dual to (9.34). Hence, by the Monge–Kantorovich theorem we have the following result.

THEOREM 9.27. *The upper arbitrage bound coincides with the price of the cheapest superreplicating portfolio.*

Assume that the bank prices the option at a price greater than $V^{\max}$. Then there is a superreplicating portfolio at price $V^{\max}$; thus an investor can obtain, at a lower price than the option, a security that yields a payoff which is always greater than the option price. Hence, no rational investor will purchase the option. Therefore, to be consistent with the absence of arbitrage, the price of the option should be less than or equal to $V^{\max}$.

9.8 REFERENCES AND NOTES

This section is based on [66]; see also references therein.

—10—

Conclusion

To conclude these notes, let us make some remarks about the nature of the mathematical methods of computation, duality, and equilibrium that have been used, and which have been recurrent themes throughout this book.

10.1 MATHEMATICS

Let us start with a discussion of the mathematical methods used in this book, as it is always good to know on which foundations one is walking. In economics, the existence of an equilibrium relies almost always on the existence of a fixed point, and the type of fixed point theorem used implies a lot about the type of mathematical toolbox available to solve a given problem. Mathematical economists agree that there are, more or less, three families of fixed point theorems in use in economics, and each of them is connected with a distinctive approach:

- The *metric approach*, based on Banach's fixed point theorem: a contraction mapping in a metric space has a unique fixed point.
- The *order-theoretic approach*, based on Tarski's fixed point theorem: an isotone mapping in a lattice has a nonempty set of fixed points which is a lattice.
- The *topological approach*, based on Schauder's fixed point theorem: a continuous mapping on a convex compact space has a nonempty set of fixed points which is a closed set.

Broadly speaking, convex analysis and linear programming belong to the first approach, supermodularity and isotonicity belong to the second approach, while degree theory and the existence of Nash equilibria in mixed strategies belong to the third one. From an applied perspective, the first two approaches are the most interesting, because they are constructive: iterative methods converge. Arguably, there are only two sets of efficient numerical methods to compute an equilibrium: through reformulation as a convex optimization problem (metric approach), or through reformulation as a Nash equilibrium in a supermodular game (order-theoretic approach).[1]

Optimal transport is interesting in the sense that it inherits from *both* the metric and the order-theoretic approaches, even though our presentation insisted on the metric side. It inherits from the metric approach by the formulation as a convex optimization problem and by the linear programming formulation. The distant foundational principle of most of the algorithms we have seen (either implicitly through the use of linear programming techniques, or explicitly in algorithm 5.4, which is a standard gradient descent method), is Banach's contraction mapping theorem. However, optimal transport also belongs to the order-theoretic approach through its close connection with matching theory. Canonical matching theory, as expounded in the book by Roth and Sotomayor [125], involves isotone mappings on lattices and is thus intrinsically order theoretic. Hence, lattice theory provides a wealth of algorithms to compute equilibria (stable matchings), and to begin with, the celebrated Gale and Shapley algorithm [60]. Although that algorithm does not involve money, Kelso and Crawford [86] showed that the Gale and Shapley algorithm could be extended to determine equilibrium wages. This leads to a discretized procedure which theoretically allows us to determine a stable matching up to the precision associated with the discretization. Hatfield and Milgrom [77] showed that Kelso and Crawford's algorithm (and thus, Gale and Shapley's) is simply the iteration of a monotone mapping on a lattice. Up to the notable exception of exercise 5.3, we have left the order-theoretic side of optimal transport virtually untouched in these notes. We plan to present that alternative point of view in the future.

10.2 COMPUTATION

While we have dedicated a significant share of this book to computational issues, we have in fact only scratched the surface. The numerical experiments we ran in chapters 3 and 6 gave us a sense of the limits of what we can compute using the methods advocated in these notes. For discrete problems, a standard PC can solve problems with a few thousand individuals on each side of the market within minutes. There is a sense that the main current scientific challenges with optimal transport are now computational: How can we move beyond this limit, and how can we efficiently solve problems that involve millions of individuals?

---

[1] One may object that Scarf's algorithm for computing approximate fixed points of continuous mappings belongs to the third family. However, this algorithm cannot be regarded as "computationally efficient."

The first answer lies in the use of specific algorithms. As we have argued, our choice of performing computations using a general-purpose linear programming solver such as Gurobi is convenient and didactical, but it is, however, suboptimal from the point of view of computational efficiency. Using specialized algorithms for the optimal assignment problem could improve computational times and memory requirements. In that respect, the research of a young generation of researchers such as Nicolas Boneel, Marco Cuturi, Quentin Mérigot, Jean-Marie Mirebeau, Gabriel Peyré, and Justin Solomon, among others, appears especially promising.

The second answer pertains to hardware. Obviously, problems of larger size will be handled better by machines with a faster CPU and more memory. But beyond that truism, new technological and commercial possibilities such as cloud computing or graphics processing units (GPUs) have made parallel computation very accessible to a larger public. It is even more so with the development of parallel computing capacities in scientific calculation software, such as MATLAB and R (through some dedicated packages for the latter, like `parallel` [117]). Parallel computing is extremely interesting for the type of problem we are after. In particular, many algorithms for optimal transportation belong to the Jacobi family, of which algorithm 7.9 is a prototypical example. In that algorithm, all the values of $u_{k+1}(x)$ are updated independently of each other, which means that this operation can be done in parallel. Many other instances of matching algorithms, such as Bertsekas' auction algorithm, can be parallelized. At the same time, parallelization is not a magic trick. It superimposes quite a bit of overhead time, which is the time it takes for the "manager CPU" to dispatch tasks to "workers," and thus the tasks dispatched had better be complex enough that the gains from parallelization are not offset by dispatching time. In algorithm 7.9 for instance, the updating steps are extremely simple—almost too simple in fact. This creates many interesting questions about the optimal use of parallel computation for optimal transport problems, for which we anticipate seeing much progress in the field in the years to come.

10.3 DUALITY

While the prominent role of duality in this text does need to be stressed, it has not escaped the attentive reader's notice that duality actually appears there at two levels. The first level involves the potentials u and v: it is duality between the worker's utility and the firm's surplus, or, using a slightly different reference, duality between prices and indirect utilities. Indeed, when $\Phi(x, y) = x'y$, we saw in chapter 6 that a pair of functions u and v, the solution to the dual problem, are convex conjugate, that is,

$$u(x) = \max_y \{x'y - v(y)\} \quad \text{and} \quad v(y) = \max_x \{x'y - u(x)\}.$$

Beyond the scalar-product surplus, chapter 7 has demonstrated that most of those ideas can be extended, and that convex conjugacy can be replaced by the more general notion of Φ-convex conjugacy, in such a way that

$$u(x) = \max_y \{\Phi(x, y) - v(y)\} \quad \text{and} \quad v(y) = \max_x \{\Phi(x, y) - u(x)\}, \quad (10.1)$$

and most results remain valid.

The second level is perhaps less obvious. It is apparent at various places in these notes (more especially in section 9.2), that minus the value of the Monge–Kantorovich problem can be written as

$$\max_{w \in \mathbb{R}^{\mathcal{Y}}} \left\{ \sum_{y \in \mathcal{Y}} w_y q_y - W(w) \right\},$$

where

$$W(w) = \mathbb{E}_P \left[\max_y \{\Phi_{Xy} + w_y\} \right]$$

is called the social welfare function in the literature. Hence, the value of the Monge–Kantorovich problem is $-W^*(q)$, where W^* is the Legendre–Fenchel transform of W. We therefore discover a second layer of duality, which is now between the social welfare function and the value of the Monge–Kantorovich problem. This duality is the duality between the Roy selection model, expressed by W, where the prices (w_y) are exogenous and the market shares (q_y) are endogenous, and the Monge–Kantorovich matching model, expressed by W^*, where the market shares (q_y) are exogenous but the prices (w_y) are now endogenous. Nothing summarizes this duality better than the subdifferential inversion formula,

$$q \in \partial W(w) \quad \Leftrightarrow \quad w \in \partial W^*(q);$$

the left-hand side consists of the determination of the market shares q given prices, while the right-hand side consists of the determination of the prices in order to adjust the market shares q. Recognizing this structure is quite illuminating; actually, several of the applications presented in chapter 9 are related to it.

The occurrence of convex duality at this higher level is one of the pleasant surprises of optimal transport theory: the investment made when studying convex analysis pays off twice! It is useful in the first place for the qualitative description of the properties of the dual solutions, and a second time to understand the structure of the Monge–Kantorovich problem itself, and in particular, the mathematical link between the optimal assignment model and the selection model.

10.4 TOWARD A THEORY OF "EQUILIBRIUM TRANSPORT"

Let us express a regret before concluding. These notes have underemphasized the importance of equilibrium, which is somewhat paradoxical for a text focused on economic applications. Actually, there is a solid reason for doing so: the Monge–Kantorovich theorem implies that equilibrium is equivalent to optimality. Given the fact that there are so many powerful optimization tools, and primarily linear programming, we have taken advantage of this fact to focus on the optimization aspect.

However, in many realistic settings, one is led to problems that are intrinsically *not* optimization problems. In the worker–firms example we began with, the expression of the firm's surplus as

$$v(y) = \max_x \{\Phi(x, y) - u(x)\}$$

relies on the assumption that the firm pays no tax on the worker's wage. In practice, firms pay taxes, which often depend both on the type of the firm and on the type of the employee. For instance, firms of certain industries may be tax exempt for hiring workers in certain age categories. Therefore, the firm's problem becomes

$$v(y) = \max_x \{\Phi(x, y) - (1 + \tau(x, y)) u(x)\},$$

where $\tau(x, y)$ is the tax on the wage levied on a firm of type y hiring a worker of type x. More generally, one may assume

$$v(y) = \max_x \mathcal{V}_{xy} (u(x)), \qquad (10.2)$$

where $u \to \mathcal{V}_{xy}(u)$ is assumed strictly decreasing and continuous. Letting $\mathcal{U}_{xy} = \mathcal{V}_{xy}^{-1}$, it is not hard to see that worker x's problem is

$$u(x) = \max_y \mathcal{U}_{xy} (v(y)); \qquad (10.3)$$

thus relations (10.2), (10.3) provide a generalization of the conjugacy of Φ-convex functions, which is called a *Galois connection*, studied in the framework of abstract convexity theory. In this setting, one may formulate the problem of equilibrium matching. Let $\Psi_{xy}(u, v) = v - \mathcal{V}_{xy}(u)$, and note that $\Psi_{xy}(u, v)$ is continuous and strictly increasing in both its arguments. We are then able to make the following definition.

DEFINITION 10.1 (Equilibrium transport problem). *Define* (π, u, v) *as a solution to the equilibrium transport problem if*

(i) $\pi \in \mathcal{M}(P, Q)$;
(ii) $\Psi_{xy} (u(x), v(y)) \geq 0 \ \forall x \in \mathcal{X}, \ y \in \mathcal{Y}$;
(iii) $\Psi_{xy} (u(x), v(y)) = 0$ *holds* π-*almost surely.*

This type of problem is studied in economics in the discrete case under the name of matching with imperfectly transferable utility. However, a systematic investigation of equilibrium transport in a more general setting remains to be conducted, despite some recent and promising advances, such as the work by Trudinger on prescribed Jacobian equations. Given the development of the applications in economics and other fields, the need for an equivalent of the wonderful Villani [148] book for equilibrium (as opposed to optimal) transport problems can be felt. We will therefore leave it as an open problem, which will mark the end of this text.

OPEN PROBLEM. *Under what conditions on Ψ, P, and Q does the equilibrium transport problem have solutions?*

10.5 REFERENCES AND NOTES

For a discussion on the three families of fixed point theorems, see the lecture notes by Ok [114, chapter 6]. Section 10.1 mentions the treatise by Roth and Sotomayor [125], who emphasize the lattice approach to matching. Crawford and Knoer [41] and Kelso and Crawford [86] extend the Gale and Shapley [60] algorithm in order to allow for a finite set of salaries. Hatfield and Milgrom [77] show that the algorithm by Kelso and Crawford can be interpreted as the iteration of an isotone mapping, which leads to the existence and the lattice structure of stable matchings in a large class of problems. In section 10.2, Bertsekas' algorithm is mentioned; the algorithm is described in Bertsekas' original papers [16, 17], and in many of his textbooks, including [18]. A reference for the "higher level" of duality mentioned in section 10.3 is [40], as well as the TraME software [70], which systematically exploits this idea. Section 10.4 mentions Galois connections; a mathematical reference is the textbook by Singer [139]. This theory is applied to implementability problems in mechanism design in Noldeke and Samuelson [113], and to discrete choice problems in Bonnet, Galichon, and Shum [20]. There is a growing literature on matching with Cobb–Douglas (ITU). An algorithmic approach based on the discretization of the set of possible transfers is the route taken by Kelso and Crawford [86], and Hatfield and Milgrom [77]. Another stream of the literature is in the tradition of cooperative games: see Kaneko [81], Gale [59], and Alkan and Gale [2]. A recent preprint by Galichon, Kominers, and Weber [67] provides an empirical framework for matching models with Cobb–Douglas. The TraME software [70] offers several computational algorithms for matching with Cobb–Douglas. The reference to Trudinger's paper is [145].

Appendix A

Solutions to the Exercises

A.1 SOLUTIONS FOR CHAPTER 2

SOLUTION TO EXERCISE 2.1. If $\pi \in \mathcal{M}(P, Q)$, then

$$\int (\Phi(x, y) + a(x) + b(y))\, d\pi(x, y)$$

$$= \int \Phi(x, y)\, d\pi(x, y) + \int a(x)\, dP(x) + \int b(y)\, dQ(y),$$

while if $u(x) + v(y) \geq \Phi(x, y)$, then

$$u(x) + a(x) + v(y) + b(y) \geq \Phi(x, y) + a(x) + b(y);$$

thus the primal solutions are unchanged, and the new dual variables are equal to the previous dual variables $u(x)$ and $v(y)$ plus $a(x)$ and $b(y)$ respectively. □

SOLUTION TO EXERCISE 2.2. One uses the fact that $x'y = \frac{1}{2}\|x\|^2 + \frac{1}{2}\|y\|^2 - \frac{1}{2}\|x - y\|^2$; thus, setting $a(x) = \frac{1}{2}\|x\|^2$ and $b(y) = \frac{1}{2}\|y\|^2$ and making use of exercise 2.1, one finds that $B = \int a(x)\, dP(x) + \int b(y)\, dQ(y) - W$. The dual for problem (2.10) is

$$\min_{u,v} \mathbb{E}_P\left[u(x)\right] + \mathbb{E}_Q\left[v(x)\right]$$

$$\text{s.t. } u(x) + v(y) \geq x'y,$$

while the dual for problem (2.11) is

$$\max_{u,v} \mathbb{E}_P\left[u(x)\right] + \mathbb{E}_Q\left[v(x)\right]$$

$$\text{s.t. } u(x) + v(y) \leq \frac{1}{2}\|x - y\|^2.$$

If (u, v) is a pair of solutions to problem (2.10), then $\frac{1}{2}\|x\|^2 - u(x)$ and $\frac{1}{2}\|y\|^2 - v(y)$ form a pair of solutions to problem (2.11). □

SOLUTION TO EXERCISE 2.3. Let us write the minimax formulation for problem (2.12). One has

$$\inf_{\substack{u,v \\ \lambda \in \mathbb{R}}} \max_{\pi \geq 0} \mathbb{E}_Q\left[v(Y)\right] + \lambda\mathbb{E}_P\left[u(X)\right] + \int_{\mathcal{X}\times\mathcal{Y}} \Phi(x,y) - u(x) - v(y)\, d\pi(x,y),$$

so, admitting the minimax principle, this is equal to

$$\max_{\substack{\pi \geq 0 \\ \lambda \in \mathbb{R}}} \int \Phi(x,y)\, d\pi(x,y)$$

$$\text{s.t. } \int_{\mathcal{X}\times\mathcal{Y}} u(x)\, d\pi(x,y) = \lambda\mathbb{E}_P\left[u(X)\right] \quad \forall u,$$

$$\int_{\mathcal{X}\times\mathcal{Y}} v(y)\, d\pi(x,y) = \mathbb{E}_Q\left[v(Y)\right] \quad \forall v.$$

Setting $v = 1$ in the second constraint implies $\int_{\mathcal{X}\times\mathcal{Y}} d\pi(x,y) = 1$; setting $u = 1$ in the second constraint implies $\int_{\mathcal{X}\times\mathcal{Y}} u(x)\, d\pi(x,y) = \lambda$; thus $\lambda = 1$, and the problem coincides with the classical Monge–Kantorovich problem. The shadow price of the bottom constraint is 1; hence, if $\mathbb{E}_Q\left[v(Y)\right]$ is interpreted as the total payoff of firms, and $\mathbb{E}_P\left[u(X)\right]$ the total payoff of workers, this means that one can substitute the payoff of firms for the payoff of workers at the rate of one for one. □

SOLUTION TO EXERCISE 2.4. When $\Phi(x,y) = a(x) + b(y)$, for any $\pi \in \mathcal{M}(P,Q)$ one has $\mathbb{E}_\pi\left[\Phi(X,Y)\right] = \mathbb{E}_P\left[a(X)\right] + \mathbb{E}_Q\left[b(Y)\right]$; hence all the $\pi \in \mathcal{M}(P,Q)$ are solutions to the primal problem. The solutions of the dual are of the form $u(x) = a(x) + c$, $v(y) = b(y) - c$. This corresponds to an economic situation where there is no complementarity between firm and worker: there is a worker fixed effect $a(x) + c$ and a firm fixed $b(y) - c$, and if firm x and worker y match, they generate the sum of their individual effects, that is, $a(x) + b(y)$. □

SOLUTION TO EXERCISE 2.5.   (i) Whatever the wage $w(x,y)$ is, the joint surplus generated by a pair (x,y) is $\alpha(x,y) + w(x,y) + \gamma(x,y) - w(x,y)$, that is, $\alpha(x,y) + \gamma(x,y) = \Phi(x,y)$. Hence the optimal assignment maximizes $\mathbb{E}_\pi\left[\Phi(X,Y)\right]$ subject to $\pi \in \mathcal{M}(P,Q)$.

(ii) Assume that $w(x,y)$ is an equilibrium wage, associated with the assignment $\pi \in \mathcal{M}(P,Q)$. Then, letting

$$u(x) = \max_y \{\alpha(x,y) + w(x,y)\} \tag{A.1}$$

be the indirect surplus of the workers, this implies $u(x) \geq \alpha(x,y) + w(x,y)$, with equality on the support of π. Similarly, letting

$$v(y) = \max_x \{\gamma(x,y) - w(x,y)\}, \tag{A.2}$$

this implies $v(y) \geq \gamma(x, y) - w(x, y)$ with equality on the support of π. As a result,

$$\gamma(x, y) - v(y) \leq w(x, y) \leq u(x) - \alpha(x, y) \quad \text{(A.3)}$$

with equality on the support of π. Thus

$$\alpha(x, y) + \gamma(x, y) \leq u(x) + v(y) \quad \text{(A.4)}$$

with equality on the support of π, therefore (u, v) is a solution to the dual problem. Conversely, if (u, v) is a solution to the dual problem, then there is a wage function $w(x, y)$ such that (A.3) holds with equality on the support of π; thus u, v, and w are related by (A.1) and (A.2); thus w is interpreted as a Walrasian wage. □

SOLUTION TO EXERCISE 2.6. Let x be the house characteristics, and y the tenant characteristics. Assume that tenant y's utility from renting house x is $S(x, y)$, and let $F(x)$ be the fee associated with house x. Let $w(x)$ be the market rent of house x. Before the bill is voted for, renter y's surplus of renting house x was

$$v(y) = \max_x \{S(x, y) - F(x) - w(x)\},$$

while the owner of house x's surplus was simply

$$u(x) = w(x).$$

Therefore, (u, v) is the solution to the dual Monge–Kantorovich problem with surplus $\Phi(x, y) = S(x, y) - F(x)$ such that $\min\{u(x) : x \in \mathrm{Supp}\,(P)\} = 0$.

After the bill is passed, renter y's surplus is now

$$v(y) = \max_x \{S(x, y) - w(x)\}$$

while the owner of house x's surplus is

$$u(x) = w(x) - F(x).$$

Therefore, (u, v) is still the solution to the dual Monge–Kantorovich problem with surplus $\Phi(x, y) = S(x, y) - F(x)$ such that $\min\{u(x) : x \in \mathrm{Supp}\,(P)\} = 0$, and the market rent $w(x)$ has just been increased by $F(x)$. Therefore the broker fee is nominally paid by the landlords, but fully reflected in the rent. The market equilibrium has fully offset the policy intervention. □

A.2 SOLUTIONS FOR CHAPTER 3

SOLUTION TO EXERCISE 3.1. Let $w = \begin{pmatrix} \text{vec}(u) \\ \text{vec}(v) \end{pmatrix}$, so that the objective function $d'w$ is $\sum_x p_x u_x + \sum_y q_y v_y$, and

$$A' = \left(\left(1'_M \otimes I_N \right)', \left(I_M \otimes 1'_N \right)' \right)$$
$$= \left(\left(1_M \otimes I_N \right), \left(I_M \otimes 1_N \right) \right).$$

Hence $A'w = (1_M \otimes I_N)\text{vec}(u) + (I_M \otimes 1_N)\text{vec}(v) = \text{vec}\left(u1'_M + 1_N v' \right)$; hence, $A'w \geq c$ reads $u1'_M + 1_N v' \geq \Phi$, that is, $u_x + v_y \geq \Phi_{xy}$. Then (3.17) becomes

$$\min_{u,v} \sum_x p_x u_x + \sum_y q_y v_y,$$
$$u_x + v_y \geq \Phi_{xy}.$$

See the codes in the online material [62]. □

SOLUTION TO EXERCISE 3.2.   (i) Without loss of generality, one may assume $\max_x u_x = 1$. Hence $v_y = \max_x \{u_x - c_{xy}\} \leq 1$, and $v_y \geq \max_x u_x - \max_x c_{xy} \geq 0$. Finally, $u_x = \min_y \{c_{xy} + v_y\} \geq 0$.

(ii) $\int_0^1 \left(1\{t \leq u_x\}, 1\{t \leq v_y\} \right) dt = \left(\int_0^{u_x} dt, \int_0^{v_y} dt \right) = (u_x, v_y)$.

(iii) If $u_x^t = 0$ or $v_y^t = 1$, then the constraint is trivially satisfied. Assume $u_x^t = 1$ and $v_y^t = 0$. Then $t \leq u_x$ and $v_y < t$, thus $u_x - v_y > 0$, hence $c_{xy} > 0$, hence $c_{xy} = 1$, and thus $u_x^t - v_y^t = c_{xy}$, hence the constraint is satisfied in all cases.

(iv) Consider a solution (u, v) to problem (3.18). As (u, v) is a convex combination of the vectors (u^t, v^t) which satisfy the constraint, it follows that any of the (u^t, v^t) is also a solution. This solution turns out to have all entries valued in $\{0, 1\}$. □

SOLUTION TO EXERCISE 3.3. Assume that Π^* is a solution to (3.10), with surplus Φ. Consider an auxiliary surplus $\tilde{S}$ whose entries are i.i.d. $\mathcal{U}([0, 1])$ random variables. There is zero probability that total surplus associated to two different permutations will coincide. Hence the solution to the linear programming problem

$$\max_{\Pi \geq 0} \sum_{xy} \Pi_{xy} \tilde{S}_{xy}$$
$$\text{s.t. } \sum_{xy} \Pi_{xy} \Phi_{xy} = \sum_{xy} \Pi_{xy}^* \Phi_{xy},$$
$$\sum_{y=1}^n \Pi_{xy} = 1, \quad \text{and} \quad \sum_{x=1}^n \Pi_{xy} = 1,$$

is integral with probability 1, and is also a solution to (3.10). Letting $z^* = \text{vec}(\Pi^*)$, this problem is computed by

$$\max_{z \geq 0} \text{vec}\left(\tilde{S}\right)' z$$

$$\text{s.t. } \left(1'_N \otimes I_N\right) z = 1_N,$$

$$\left(I_N \otimes 1'_N\right) z = 1_N,$$

$$\text{vec}\left(\Phi\right)' z = \text{vec}\left(\Phi\right)' z^*.$$

$\square$

SOLUTION TO EXERCISE 3.4. One sees that the set obtained is the frontier of a convex set. Indeed, the set

$$C = \left\{\left(\mathbb{E}_\pi\left[X_1 X_2\right], \mathbb{E}_\pi\left[Y_1 Y_2\right]\right) \in \mathbb{R}^2 : \pi \in \mathcal{M}(P, Q)\right\}$$

is a convex set, and C_θ is the maximizer of $C_1 \cos\theta + C_2 \sin\theta$ over $C = (C_1, C_2) \in \mathbb{R}^2$; hence it lies on the frontier of C. $\square$

SOLUTION TO EXERCISE 3.5. The dual problem is written as

$$\min_{u, v \geq 0} \sum_{x=1}^{N} p_x u_x + \sum_{y=1}^{M} q_y v_y$$

$$\text{s.t. } u_x + v_y \geq \Phi_{xy}.$$

Compared with problem (3.4), the only modification is that the dual variables u and v are assumed nonnegative. This can be interpreted as agents having an outside option of value 0 if they choose not to match. Therefore, in addition to the constraint expressing the absence of blocking pair $u_x + v_y \geq \Phi_{xy}$, the constraints $u_x \geq 0$ and $v_y \geq 0$ express that a single individual may be blocking. $\square$

SOLUTION TO EXERCISE 3.6. Recall that, if $\left(u_i, v_j\right)$ are the solution to the dual problem, then

$$u_i = \max_{j \in \mathcal{J}}\left\{\Phi_{ij} - v_j, 0\right\} \quad \text{and} \quad v_j = \max_{i \in \mathcal{I}}\left\{\Phi_{ij} - u_i, 0\right\}.$$

Hence, $\Phi_{ij} = \Phi_{x_i y_j}$ implies $u_i = u_{x_i}$, where $u_x = \max_{j \in \mathcal{J}}\left\{\Phi_{xy_j} - v_j, 0\right\}$ and similarly, $v_j = v_{y_j}$, where $v_y = \max_{i \in \mathcal{I}}\left\{\Phi_{x_i y} - u_i, 0\right\}$. $\square$

A.3 SOLUTIONS FOR CHAPTER 4

SOLUTION TO EXERCISE 4.1. Take $d = 2$, and assume that X is uniformly distributed over $[0, 1]^2$. Then X_1 and X_2 are independent, $F_P(X) = X_1 X_2$, and for $a \in (0, 1)$, one has

$$\Pr(F_P(X) \le a) = \mathbb{E}\left[1\{X_1 X_2 \le a\}\right] = \mathbb{E}\left[\min(a/X_1, 1)\right] = a(1 - \ln a) > a. \qquad \square$$

SOLUTION TO EXERCISE 4.2. (i) Let x_{-1} be the lower bound of the support of P (possibly $-\infty$), and define x_j, $1 \le j \le M$ such that

$$x_j = \sum_{k \le j} q_k.$$

By theorem 4.3, the optimal coupling (X, Y) has $Y = T(X)$, where T is such that $T(x) = y_j$ if $x \in (x_{j-1}, x_j)$. In particular, we may take

$$T(x) = \sum_{j=0}^{M} y_j 1\left\{x \in [x_{j-1}, x_j)\right\}.$$

Then

$$W = \sum_{j=0}^{K} y_j \mathbb{E}\left[X 1\left\{X \in [x_{j-1}, x_j)\right\}\right] = \sum_{j=0}^{K} y_j (x_j - x_{j-1})\left(\frac{x_{j-1} + x_j}{2}\right).$$

Thus,

$$W = \sum_{j=0}^{K} y_j q_j \sum_{k \le j-1} q_k + \frac{q_j}{2} = \sum_{k < j} y_j q_j q_k + \frac{1}{2} \sum_{j=1}^{K} y_j q_j^2,$$

so

$$W = \frac{1}{2} \sum_{k,j} \max(y_k, y_j) q_j q_k$$

is a quadratic expression in q.

(ii) We now turn to the expression of the dual of the Monge–Kantorovich problem, in order to compute the wages $u(x)$ and surpluses v_j. We normalize v_0 such that $v_0 = 0$, which implies $u(0) = 0$. Write the

Monge–Kantorovich dual

$$W = \min_{u(x), v_j} \int u(x)dP(x) + \sum_{j=1}^{M} q_j v_j$$

$$\text{s.t. } u(x) + v_j \geq xy_j \quad \forall x, \forall j,$$

$$v_0 = 0;$$

hence, by the envelope theorem,

$$v_j = \frac{\partial W}{\partial q_j}. \tag{A.5}$$

Because of the normalization $v_0 = 0$, we have to rewrite the expression for W by substituting out q_0 and replacing it with $1 - \sum_{j \neq 0} q_j$. One has

$$W = \frac{1}{2} \sum_{k,j \neq 0} A_{jk} q_j q_k + \sum_{k \neq 0} b_k q_k + c,$$

where

$$A_{jk} = \max\left(y_k, y_j\right) - \max\left(y_k, y_0\right) - \max\left(y_j, y_0\right) + y_0,$$

$$b_k = \max\left(y_k, y_0\right) - y_0,$$

$$c = \tfrac{1}{2} y_0.$$

By expression (A.5) in the solution to exercise 4.2, this yields an expression for v:

$$v_j = \sum_{k \neq 0} A_{jk} q_k + b_j;$$

hence

$$u(x) = \max_{j=0,\dots,M} \left\{xy_j - v_j\right\}$$

is a convex and piecewise affine function. □

SOLUTION TO EXERCISE 4.3. See the codes in the online material [62]. □

SOLUTION TO EXERCISE 4.4. Let $W = u(X)$. One has

$$u(x) = \int_0^x at^{a-1} Q(t)^b \, dt,$$

where Q is the quantile of the standard normal distribution. The c.d.f. of W is $F_W(w) = \Pr(u(X) \le w) = \Pr\left(X \le u^{-1}(w)\right) = u^{-1}(w)$. The Gini index is given by

$$\frac{1}{\mathbb{E}[W]} \int_0^\infty F_W(w)(1 - F_W(w))\, dw.$$

See the codes in the online material [62]. □

SOLUTION TO EXERCISE 4.5.   (i) The new surplus function is still supermodular, and workers are still in excess supply, so the matching patterns are unchanged. Let x_0 be the lowest talent of the employed workers. By the differential wage equation,

$$u'(x) = \partial_x \Phi(x, T(x)) + f'(x);$$

hence, after the shock the worker of productivity x_0 still gets 0, and integrating the differential wage equation,

$$u(x) = \int_{x_0}^x \partial_x \Phi(z, T(z))\, dz + \int_{x_0}^x f'(z)\, dz;$$

hence, the differential in the salary of employed workers of type x is $f(x) - f(x_0)$.

The effect on the salaries of the workers is ambiguous:

- If f is increasing, then there is a general rise in the salaries.
- If f is decreasing, then there is a general decrease in the salaries.

The last result may appear counterintuitive, but a decreasing f will decrease the differentiation among the worker's productivity, and increase local competition, and thus decrease the rent of the workers. In the case of increasing f, the differentiation among workers increases and so does the rent extracted from the sorting process. They are able to capture a part of the value creation, but not all of it due to the fact that they are in excess supply.

(ii) The differential wage equation is now written

$$u'(x) = \partial_x \Phi(x, T(x)),$$

which integrates into

$$u(x) = \int_{x_0}^x \partial_x \Phi(z, T(z))\, dz.$$

In this case, the compensation of the workers is unchanged: all the production shock has been captured by the firm. This is due to the fact that firms are in short supply, so they are able to capture the totality of the extra value created. As this extra value creation is homogeneous across workers, they do not need to pay them an extra sorting rent. □

SOLUTION TO EXERCISE 4.6. The fiscal gain from marriage is the difference between the after-tax income of the household and the sum of the after-tax income of the partners if they were to remain single. The after-tax income of the household is $x + y - 2\tau\left(\frac{1}{2}(x + y)\right)$, while the after tax income of x and y if single is respectively $x - \tau(x)$ and $y - \tau(y)$. Thus the fiscal gain from marriage is $\Phi(x, y) = \tau(x) + \tau(y) - 2\tau\left(\frac{1}{2}(x + y)\right)$, which is positive if the tax schedule is convex. Further, this surplus function exhibits

$$\frac{\partial^2 \Phi(x, y)}{\partial x\, \partial y} \leq 0,$$

which is *negative assortative matching*: as a result, this theory predicts that in an equilibrium, the poorest man will marry the richest woman, and the richest man will marry the poorest woman—a somewhat counterfactual prediction! □

A.4 SOLUTIONS FOR CHAPTER 5

SOLUTION TO EXERCISE 5.1. A power diagram is left unchanged if one adds the same constant to all v_j values. Hence, without loss of generality, one may assume $w_j \geq 0$. For each j, introduce

$$\tilde{y}_j = \left(y_j, \sqrt{2w_j}\right) \in \mathbb{R}^{d+1}.$$

The Voronoi diagram associated with $\tilde{y}_j$ is such that $\tilde{x}$ is in the cell of $\tilde{y}_j$ if and only if

$$\left|\tilde{x} - \tilde{y}_j\right|^2 \leq \left|\tilde{x} - \tilde{y}_k\right|^2$$

for all k; that is,

$$2\tilde{x} \cdot \tilde{y}_j - \left|\tilde{y}_j\right|^2 \geq 2\tilde{x} \cdot \tilde{y}_k - \left|\tilde{y}_k\right|^2\,;$$

that is,

$$\tilde{x} \cdot \left(\tilde{y}_j - \tilde{y}_k\right) \geq \frac{\left|\tilde{y}_j\right|^2 - \left|\tilde{y}_k\right|^2}{2}\,;$$

that is, if $\tilde{x} = (x, 0)$,

$$x \cdot (y_j - y_k) \geq \frac{|y_j|^2}{2} + w_j - \frac{|y_k|^2}{2} - w_k;$$

thus

$$-\frac{|x - y_j|^2}{2} - w_j \geq -\frac{|x - y_k|^2}{2} - w_k.$$

Hence, one sees that the power diagram in $\mathbb{R}^d$, associated with the system (y_j, w_j), coincides with the intersection of the Voronoi diagram in $\mathbb{R}^{d+1}$, associated with the $\tilde{y}_j$ values, with the hyperplane $x_{d+1} = 0$. □

SOLUTION TO EXERCISE 5.2. In this case,

$$F(v) = \int \max_j \{x \cdot y_j - v_j\} \, dP(x) + \sum_{j=1}^{M} q_j v_j$$

has an explicit formula. Indeed, by the definition of y_j,

$$x \cdot y_j = x_j;$$

thus

$$\int \max_j \{x \cdot y_j - v_j\} \, dP(x) = \mathbb{E}\left[\max_j \{-v_j + X_j\} \right].$$

One of the defining properties of the Gumbel distribution is that

$$\mathbb{E}\left[\max_j \{-v_j + X_j\} \right] = \log \sum_{j=1}^{m} e^{-v_j},$$

thus

$$F(v) = \log \sum_{j=1}^{m} e^{-v_j} + \sum_{j=1}^{M} q_j v_j;$$

so by first-order conditions,

$$q_j = \frac{e^{-v_j}}{\sum_{j=1}^{m} e^{-v_j}},$$

hence

$$v_j = -\ln q_j + c,$$

where c is any constant. The value of the social planner's program is then F evaluated at that v, that is,

$$-\sum_{j=1}^{m} q_j \ln q_j,$$

which is the entropy of q. $\qquad\square$

SOLUTION TO EXERCISE 5.3. (i) For $l \neq k$, one has

$$\frac{\partial^2 \Pi}{\partial v_k \, \partial v_l} (v_k; v_{-k}) = -\frac{\partial^2 W}{\partial v_k \, \partial v_l}$$

$$= \frac{\partial}{\partial v_l} \mathbb{E}_P \left[1 \left\{ X' y_k - v_k \geq \max_{j \neq k} X' y_j - v_j \right\} \right]$$

$$= \mathbb{E}_P \left[\delta \left(X' (y_k - y_l) - (v_k - v_l) \right) 1 \left\{ X' y_l - v_l \geq \max_{j \neq k,l} X' y_j - v_j \right\} \right];$$

thus $\partial^2 \Pi(v_k; v_{-k})/\partial v_k \, \partial v_l \geq 0$, and W is clearly convex with respect to v_k, hence Π_k is concave in v_k. Hence the game is supermodular, and as a result, if $V_k(v_{-k})$ is the best reply of player k, V_k is isotone, which means that $v_{-k} \leq v'_{-k}$ implies $V_k(v_{-k}) \leq V_k(v'_{-k})$. This result is known as Topkis' theorem and can be seen in the differentiable case by noting that $\partial \Pi(V_k(v_{-k}); v_{-k})/\partial v_k = 0$; thus

$$\frac{\partial V_k(v_{-k})}{\partial v_l} = -\frac{\partial^2 \Pi/\partial v_k \partial v_l}{\partial^2 \Pi/\partial v_k^2} \geq 0.$$

(ii) Let $\bar{v}_k$ be such that $\Pr\left(X' y_1 \geq X' y_k - v_k \right) \geq 1 - q_k$. Then $\bar{v}_k \geq v_k^*$. As a result, the dynamics provided by (5.23), which is simply the best response dynamics $v_k^{t+1} = V_k\left(v_{-k}^t \right)$, converges toward v^*, which is a fixed point of V. The best response dynamics is interesting in the sense that (a) it only involves iterated single-dimensional optimization operations and (b) it is naturally parallelizable, as the $V_k(v_{-k}^t)$ can be computed in parallel. $\qquad\square$

SOLUTION TO EXERCISE 5.4. (i) Consider a solution (y, q) and π the associated optimal transportation plan. Then

$$W(y, q) = \sum_{k=1}^{n} q_k \mathbb{E}_\pi [(X - y_k)^2 | Y = y_k];$$

hence, optimality with respect to y implies

$$y_k = \mathbb{E}_\pi [X | Y = y_k].$$

(ii) The following procedure uses the fact that the y values have to be the barycenter of the corresponding Voronoi cells.

Start with an initial guess of $y_1^0, \ldots, y_n^0$.
At step t, compute the Voronoi cells $V_k^t = \{x : |x - y_k^t| \leq |x - y_j| \; \forall j\}$, and then update the values of y by the values of the barycenter of the cells, that is,

$$y_k^{t+1} = \frac{1}{P(V_k^t)} \int_{V_k^t} x \, dP(x).$$

Stop when the y^{t+1} values are close enough to the y^t values.
This algorithm is known as Lloyd's algorithm and was first described in [95]. $\quad\square$

SOLUTION TO EXERCISE 5.5. Recall that $\Pr(\mathcal{X}_k^v) = -\partial W/\partial v_k$, where W is defined by (5.10). The profit maximization condition for fountain k implies the first-order conditions for equilibrium

$$\frac{\partial W}{\partial v_k} + v_k \frac{\partial^2 W}{\partial v_k^2} = 0.$$

In the setting of exercise 5.2,

$$\Pr\left(\mathcal{X}_k^v\right) = \frac{e^{-v_k}}{\sum_{j=1}^M e^{-v_j}},$$

therefore first-order conditions yield

$$\frac{1}{v_k} - 1 + \frac{e^{-v_k}}{\sum_{j=1}^M e^{-v_j}} = 0.$$

The symmetric equilibrium satisfies $v_k = v$, where

$$\frac{1}{v} - 1 + \frac{e^{-v}}{M e^{-v}} = 0,$$

thus

$$v = \frac{M}{M - 1}.$$

$\square$

SOLUTION TO EXERCISE 5.6. See the codes in the online material [62]. $\quad\square$

A.5 SOLUTIONS FOR CHAPTER 6

SOLUTION TO EXERCISE 6.1. (i) For $X \sim P$, we have $\nabla u(X) \sim Q$, where u is a convex function. Given R an orthogonal matrix, let $u^R(x) = u(Rx)$; u^R is convex and we have $\nabla u^R(X) = R'\nabla u(RX) \sim Q$; hence, by uniqueness of the Brenier map, $u^R(x) = u(x)$, thus $u(x) = \zeta(|x|)$.

(ii) If $u(x) = \zeta(|x|)$, then $\nabla u(x) = \zeta'(|x|)x/|x|$, and

$$D^2 u(x) = \frac{\zeta'(|x|)I_d}{|x|} + \left(\frac{\zeta''(|x|)}{|x|^2} - \frac{\zeta'(|x|)}{|x|^3} \right) xx';$$

note that $aI_d + bxx'$ has eigenvalues a with multiplicity $d - 1$, and $a + b|x|^2$ with multiplicity 1. Thus the eigenvalues of $D^2 u(x)$ are positive if $\zeta'(|x|) > 0$ and $\zeta''(|x|) \geq 0$. Hence, $x \to \zeta(|x|)$ is convex if and only if ζ is nondecreasing and convex.

(iii) One has $|Y| = |\nabla(X)| = \zeta'(|X|)$, thus ζ' is an increasing map such that the distribution of $\zeta'(|X|)$ is the distribution of $|Y|$. Hence,

$$\zeta(t) = F_{|Y|}^{-1} \circ F_{|X|}.$$

□

SOLUTION TO EXERCISE 6.2. The following is just a rough sketch of the argument; see details in [34]. Let $X = (X_1, X_2) \sim P$, and for $\lambda > 0$, let $Y^\lambda = (Y_1^\lambda, Y_2^\lambda) \sim Q$ be such that the distribution of (X, Y^λ) is optimal for

$$\max_{\pi \in \mathcal{M}(P,Q)} \mathbb{E}_\pi [X_1 Y_1 + \lambda X_2 Y_2].$$

Let $Y^* = \lim_{\lambda \to 0} Y^\lambda$, and let $\bar{Y} = \bar{T}(X)$. We would like to show that $Y^* = \bar{Y}$. Assume that Y_1^* is not an increasing function of X_1. Then one has for some $\epsilon > 0$,

$$\mathbb{E}\left[X_1 Y_1^* \right] + 2\epsilon \leq \mathbb{E}\left[X_1 \bar{Y}_1 \right];$$

thus for $\lambda < \epsilon / \sqrt{\mathbb{E}\left[X_1^2 \right] \mathbb{E}\left[Y_1^2 \right]}$,

$$\mathbb{E}_\pi \left[X_1 Y_1^\lambda + \lambda X_2 Y_2^\lambda \right] \leq \mathbb{E}_\pi \left[X_1 \bar{Y}_1 + \lambda X_2 \bar{Y}_2 \right],$$

a contradiction. Thus Y_1^* is an increasing function of X_1, and $Y_1^* = \bar{Y}_1$. Next, let us argue that $Y_2^* = \bar{Y}_2$. Assume

$$\mathbb{E}\left[X_2 Y_2^* \right] \neq \max_{\tilde{Y}_2 : (\bar{Y}_1, \tilde{Y}_2) \sim Q} \mathbb{E}\left[X_2 \tilde{Y}_2 \right]; \tag{A.6}$$

then as $\left(\bar{Y}_1, Y_2^*\right) \sim Q$, it follows that for some $\epsilon > 0$, and for λ small enough,

$$\mathbb{E}\left[X_1\bar{Y}_1 + \lambda X_2 Y_2^*\right] + 2\epsilon \leq \max_{\tilde{Y}_2:(\bar{Y}_1,\tilde{Y}_2)\sim Q} \mathbb{E}\left[X_1\bar{Y}_1 + \lambda X_2 \tilde{Y}_2\right];$$

hence, for λ small enough,

$$\mathbb{E}[X_1 Y_1^\lambda + \lambda X_2 Y_2^\lambda] + \epsilon \leq \max_{\tilde{Y}_2:(\bar{Y}_1,\tilde{Y}_2)\sim Q} \mathbb{E}[X_1 Y_1^\lambda + \lambda X_2 \tilde{Y}_2],$$

which is a contradiction, since $\mathbb{E}[X_1 Y_1^\lambda + \lambda X_2 Y_2^\lambda] = \max_{\tilde{Y}_2:(\bar{Y}_1,\tilde{Y}_2)\sim Q}$ $\mathbb{E}\left[X_1 Y_1^\lambda + \lambda X_2 \tilde{Y}_2\right]$. Hence, (A.6) is wrong, and the equality holds instead. Therefore,

$$\mathbb{E}\left[X_2 Y_2^*\right] = \max_{\tilde{Y}_2:(\bar{Y}_1,\tilde{Y}_2)\sim Q} \mathbb{E}\left[X_2 \tilde{Y}_2\right]$$

and thus, $Y_2^* = \bar{T}_2\left(X_1, X_2\right)$. □

SOLUTION TO EXERCISE 6.3. See the codes in the online material [62]. □

SOLUTION TO EXERCISE 6.4. See the codes in the online material [62]. □

SOLUTION TO EXERCISE 6.5. The primal and the dual problems are

$$\max_{\pi\in\mathcal{M}(P,Q)} \mathbb{E}_\pi\left[X'AY\right] = \min_{u(x)+v(y)\geq x'Ay} \mathbb{E}_P\left[u(X)\right] + \mathbb{E}_Q\left[v(Y)\right].$$

Let $(X, Y) \sim \pi$ be a solution to the primal, and (u, v) be a solution to the dual. Let $\tilde{Y} = AY$, $\tilde{Q} = \mathcal{N}\left(0, A\Sigma_Q A'\right)$, and $\tilde{v}(\tilde{y}) = v\left(A^{-1}\tilde{y}\right)$, so that $\tilde{Y} \sim \tilde{Q}$, and $v(y) = \tilde{v}(Ay)$. Clearly, $\left(X, \tilde{Y}\right)$ is the solution to

$$\max_{\pi\in\mathcal{M}(P,\tilde{Q})} \mathbb{E}_\pi\left[X'\tilde{Y}\right],$$

and $(u, \tilde{v})$ is a solution to

$$\min_{u(x)+\tilde{v}(\tilde{y})\geq x'\tilde{y}} \mathbb{E}_P\left[u(X)\right] + \mathbb{E}_{\tilde{Q}}\left[v(\tilde{Y})\right];$$

hence, by the results of example 6.1, we have

$$\tilde{Y} = \Sigma_P^{-1/2}(\Sigma_P^{1/2}\Sigma_{\tilde{Q}}\Sigma_P^{1/2})^{1/2}\Sigma_P^{-1/2}X,$$

thus

$$Y = A^{-1}\Sigma_P^{-1/2}(\Sigma_P^{1/2}A\Sigma_Q A'\Sigma_P^{1/2})^{1/2}\Sigma_P^{-1/2}X$$

and

$$u(x) = \frac{1}{2} x' \Sigma_P^{-1/2} (\Sigma_P^{1/2} A \Sigma_Q A' \Sigma_P^{1/2})^{1/2} \Sigma_P^{-1/2} x.$$

☐

SOLUTION TO EXERCISE 6.6. Let $T(\varepsilon)$ be the quality chosen by consumer with characteristics ε. One has $T(\varepsilon) \in \arg\max_y \{\varepsilon'y - v(y)\}$; thus, letting $u(\varepsilon) = \max_y \{\varepsilon'y - v(y)\}$, one has $T \in \partial u$. Further, one has $T\#P = Q$; thus T is the Brenier map from P to Q, and (u, v) is the solution to the dual Monge–Kantorovich problem

$$\min \left\{ \mathbb{E}_P \left[u(\varepsilon) \right] + \mathbb{E}_Q \left[v(y) \right] \text{ s.t. } u(\varepsilon) + v(y) \geq \varepsilon'y \right\}.$$

☐

A.6 SOLUTIONS FOR CHAPTER 7

SOLUTION TO EXERCISE 7.1. Let x_n be a sequence such that $\Phi(x_n, y) - f(x_n) \to f^\Phi(y)$. Then

$$\Phi(x_n, y) - f(x_n) - f^\Phi(x) \leq \Phi(x_n, y) - f(x_n) - \Phi(x_n, x) + f(x_n),$$

thus

$$\Phi(x_n, y) - f(x_n) - f^\Phi(x) \leq \Phi(x_n, y) - \Phi(x_n, x) \leq -\Phi(x, y),$$

thus

$$f^\Phi(y) - f^\Phi(x) \leq -\Phi(x, y);$$

hence,

$$\sup_{x \in \mathcal{X}} \{\Phi(x, y) - f^\Phi(x)\} \leq -f^\Phi(y).$$

But since $-f^\Phi(y) = \Phi(y, y) - f^\Phi(y)$, it follows that

$$\sup_{x \in \mathcal{X}} \{\Phi(x, y) - f^\Phi(x)\} = -f^\Phi(y);$$

hence, $f^{\Phi\Phi}(y) = -f^\Phi(y)$.

☐

Solution to exercise 7.2. Consider $\Phi(x, y) = \exp(x + y)$ and $f(x) = \frac{1}{2}\exp(2x)$. One has

$$f^\Phi(y) = \max_x \left\{\exp(x + y) - \tfrac{1}{2}\exp(2x)\right\}$$

$$= \max\left\{X\exp(y) - \tfrac{1}{2}X^2\right\} = \tfrac{1}{2}\exp(2y);$$

thus f is a Φ-convex function, and

$$\partial f^\Phi(x) = \arg\max_y \left\{\exp(x + y) - \tfrac{1}{2}\exp(2y)\right\} = x.$$

By the same token, $\partial^\Phi(2f)(x) = x - \ln 2$. Thus $\partial^\Phi(2f)(x) \neq 2\partial^\Phi f(x)$. ☐

Solution to exercise 7.3. See the parallel implementation of the IPFP in the R package TraME [70]. ☐

Solution to exercise 7.4. One fix to the problem consists of replacing the exponential function by a "modified exponential function" $\widetilde{\exp}$ such that

$$\widetilde{\exp}(t) = \begin{cases} \exp(t) & \text{if } t \leq 50, \\ (t - 49)\exp(50) & \text{otherwise,} \end{cases}$$

so that $\widetilde{\exp}$ is C^1, coincides with the exponential when $t \leq 50$, and is linear above 50. See the codes in the online material [62]. ☐

Solution to exercise 7.5. Let (u, v) be a solution to the dual problem (7.7), and let

$$u^c(x) = u(x) + c \quad \text{and} \quad v^c(y) = v(y) - c,$$

so that (u^c, v^c) is also a solution to (7.7). Then for all $x \in \mathcal{X}$ and $y \in \mathcal{Y}$,

$$u^c(x) + v^c(y) \geq \Phi(x, y) \geq a_x(x) + b_y(y),$$

thus

$$\inf_x \{u(x) - a_x(x)\} + c = \inf_x \left\{u^c(x) - a_x(x)\right\}$$

$$\geq \sup_y \left\{b_y(y) - v^c(y)\right\}$$

$$= \sup_y \left\{b_y(y) - v(y)\right\} + c,$$

and there exists a value of c such that

$$\inf_{x} \left\{ u^c(x) - a_\chi(x) \right\} \geq 0 \geq \sup_{y} \left\{ by(y) - v^c(y) \right\} ;$$

therefore $u^c(x) \geq a_\chi(x)$ for all x and $v^c(y) \geq by(y)$ for all y. □

SOLUTION TO EXERCISE 7.6. The firm's problem is

$$\min_{x} \left\{ K(x, y) W(x) \right\}$$

or equivalently, $v(y) = \max_{x} \left\{ \Phi(x, y) - u(x) \right\}$, where $u(x) = \ln W(x)$ and $\Phi(x, y) = -\ln K(x, y)$. Thus, (u, v) is the solution to

$$\min_{u, v} \int u(x) \, dP(x) + \int v(y) \, dQ(y)$$

$$\text{s.t. } u(x) + v(y) \geq \Phi(x, y),$$

and the wage $W(x)$ is obtained from $W(x) = \exp(u(x))$. □

A.7 SOLUTIONS FOR CHAPTER 8

SOLUTION TO EXERCISE 8.1.   (i) The network has a profitable loop if and only if the value of problem (8.4) is $+\infty$. This is equivalent to if and only if the dual problem is not feasible, that is, if and only if there is no potential u such that (8.6) holds.

(ii) If (x, y) and (y, x) are in $\mathcal{A}$, then $u_y - u_x \geq \Phi_{xy}$ and $u_x - u_y \geq \Phi_{yx}$; thus by summation, $\Phi_{xy} + \Phi_{yx} \leq 0$.

(iii) In this case, $u_y - u_x \geq \Phi_{xy}$ and $u_x - u_y \geq -\Phi_{xy}$; thus $u_y - u_x \geq \Phi_{xy} \geq u_y - u_x$. □

SOLUTION TO EXERCISE 8.2.   (i) We will simply give a sketch of the proof. The proof consists of showing that given an integral feasible flow π, one can write $\pi = \pi^{res} + \pi^{pl}$, where π^{pl} is either a path or a loop, and

$$\sum_{xy \in \mathcal{A}} \pi_{xy}^{res} < \sum_{xy \in \mathcal{A}} \pi_{xy}. \tag{A.7}$$

The idea consists of starting from x_0 such that $n_{x_0} < 0$, and given x_k, define x_{k+1} such that (a) $x_k x_{k+1} \in \mathcal{A}$, (b) $\pi_{x_k x_{k+1}} > 0$, (c) $n_{x_{k+1}} \leq 0$, and (d) x_{k+1} has not been visited before, until such x_{k+1} does not exist. When such x_{k+1} does not exist, then it does not exist because either (c) or (d) fails. If (c) fails, then we have identified a path from sources to destinations, and if (d) fails, then we have identified a loop. Call π^{pl} the corresponding flow, and let $\pi^{\text{res}} = \pi - \pi^{\text{pl}}$. By construction, π is a feasible flow, and we have by construction $\pi = \pi^{\text{res}} + \pi^{\text{pl}}$ and (A.7).

(ii) If π is optimal for problem (8.4), then there is no loop in the previous representation; further, all paths generate equal surplus. $\square$

SOLUTION TO EXERCISE 8.3. Let $(\mathcal{Z}, \mathcal{A}, n, \Phi)$ be a network, and assume $xx \in \mathcal{A}$ for all $x \in \mathcal{Z}$. Let $\tilde{\mathcal{Z}} = \mathcal{Z} \times \{1, \ldots, T\}$, and $\tilde{\mathcal{A}} \subseteq \tilde{\mathcal{Z}} \times \tilde{\mathcal{Z}}$ such that $(zt, z't') \in \tilde{\mathcal{A}}$ if and only if $zz' \in \mathcal{A}$ and $t' = t + 1$. Assume that $\tilde{\Phi}_{zt,z'(t+1)}$ is defined if $zz' \in \mathcal{A}$ and is such that $\tilde{\Phi}_{zt,z(t+1)} = 0$. Let $\tilde{n}_{x,1} = n_x 1\{n_x < 0\}$, $\tilde{n}_{x,T} = n_x 1\{n_x > 0\}$, and $\tilde{n}_{x,t} = 0$ for $t \in \{2, \ldots, T - 1\}$. In this case,

$$\left(\tilde{\nabla}^* \tilde{\pi}\right)_{x,t} = \sum_{y:yx \in \mathcal{A}} \tilde{\pi}_{y(t-1),xt} - \sum_{y:xy \in \mathcal{A}} \tilde{\pi}_{xt,y(t+1)};$$

thus $\tilde{\nabla}^* \tilde{\pi} = n$ implies

$$\sum_{y:xy \in \mathcal{A}} \tilde{\pi}_{x1,y2} = n_x^-,$$

$$\sum_{y:yx \in \mathcal{A}} \tilde{\pi}_{y(t-1),xt} = \sum_{y:xy \in \mathcal{A}} \tilde{\pi}_{xt,y(t+1)} \text{ for } t \in \{2, \ldots, T - 1\},$$

$$\sum_{y:yx \in \mathcal{A}} \tilde{\pi}_{y(T-1),xT} = n_x^+,$$

and, for $xy \in \mathcal{A}$,

$$(\tilde{\nabla} f)_{xt,y(t+1)} = f_{y(t+1)} - f_{xt}.$$

The time-dependent network flow is then

$$\max_{\tilde{\pi} \geq 0} \sum_{t=1}^{T-1} \sum_{xy \in \mathcal{A}} \tilde{\pi}_{xt,y(t+1)} \Phi_{xy}$$

$$\text{s.t. } \tilde{\nabla}^* \tilde{\pi} = \tilde{n},$$

which has dual

$$\min_{\tilde{v}} \sum n_x^+ \tilde{v}_{xT} - \sum n_x^- \tilde{v}_{x1}$$

$$\text{s.t. } \tilde{v}_{y(t+1)} - \tilde{v}_{xt} \geq \Phi_{xy}.$$

This setting allows for important dynamic extensions of the problems described in chapter 8. Here, we are not only concerned with going from one distribution to another, but also at what time we should be going from one to the other. □

SOLUTION TO EXERCISE 8.4. If x is a supply node, that is, if $n_x \leq 0$, then the quantity actually entering the network is $-(\nabla^*\pi)_x$; thus the constraint implies $-(\nabla^*\pi)_x \leq -n_x$, hence $(\nabla^*\pi)_x \geq n_x$. If y is a demand node, then the quantity leaving the network is $(\nabla^*\pi)_y$; thus $n_y > 0$ implies $(\nabla^*\pi)_y \leq n_y$. Letting v_y be the Lagrange multiplier of these constraints, one has $v_x \leq 0$ if x is a supply node and $v_y \geq 0$ if y is a demand node. □

SOLUTION TO EXERCISE 8.5. See the codes in the online material [62]. □

SOLUTION TO EXERCISE 8.6. See the codes in the online material [62]. □

===== Appendix B =====
Linear Programming

There are very strong links between linear programming, minimax theorems, and zero-sum games. Let us start with the statement of a minimax theorem.

B.1 MINIMAX THEOREM

There are many versions of the minimax theorem which extend von Neumann's original formulation; the present version, which we state in finite dimension for convenience, was given by Maurice Sion.

THEOREM B.1 (Minimax). *Let K be a compact convex subset of $\mathbb{R}^n$, and let C be a convex subset of $\mathbb{R}^m$. Let $f : \mathbb{R}^n \times \mathbb{R}^m \to \mathbb{R}$ be such that $f(\cdot, y)$ is convex for each y, and $f(x, \cdot)$ is concave for each x. Then*

$$\inf_{x \in K} \sup_{y \in C} f(x, y) = \sup_{y \in C} \inf_{x \in K} f(x, y).$$

B.2 DUALITY

Consider the generic linear programming (LP) problem called the *primal program*:

$$V_P = \max_x c'x$$

$$\text{s.t. } Ax = b, \qquad\qquad (\text{P})$$

$$x \geq 0.$$

This form is quite general. Indeed, if x_j is unrestricted we can introduce two variables x_j^+ and x_j^-, with $x_j^+, x_j^- \geq 0$, and set $x_j = x_j^+ - x_j^-$. We can transform additional constraints of the form $Ex \leq d$ by adding a vector of slack variables $s \geq 0$ such that $Ex + s = d$; for constraints of the form $Ex \geq d$ we subtract the slack variable $s \geq 0$: $Dx - s = d$; $\min c'x$ is replaced by $-\max c'x$; etc.

The primal program is *infeasible* if the set of x that verify the constraints is empty; in that case, we take the convention of setting its value equal to $-\infty$.

Otherwise it is *feasible*; its value is either finite if the set $\{c \cdot x : Ax = b, \ x \geq 0\}$ is bounded from above, or equal to $+\infty$ if not.

To each linear programming problem (P) one can associate another one (D) called its *dual program*:

$$V_D = \min b'y$$
$$\text{s.t. } A'y \geq c, \qquad\qquad\qquad\text{(D)}$$
$$y \text{ unrestricted,}$$

where A' is the transposed matrix of A.

Consistent with the convention taken above for the primal program, when the dual program is infeasible its value is set to $+\infty$.

One has in general the inequality

$$V_D \geq V_P,$$

which is called the *weak duality inequality*. In the case when it holds strictly, one is said to have a *duality gap*. The following result asserts that there cannot be a duality gap unless both the primal and the dual programs are infeasible.

THEOREM B.2. *Exactly one of the following statements is true:*

  (i) *Both the primal and dual programs are infeasible.*
  (ii) *Strong duality holds:*

$$V_D = V_P.$$

PROOF. Consider the primal problem

$$V_P = \max c'x$$
$$\text{s.t. } Ax = b, \qquad\qquad\qquad\text{(P)}$$
$$x \geq 0.$$

This problem can be written

$$V_P = \max_{x \geq 0} \min_y \left\{ c'x + y'(b - Ax) \right\}.$$

But the minimax inequality,

$$\max_{x \geq 0} \min_y \ \leq \ \min_y \max_{x \geq 0},$$

yields

$$V_P \leq \min_{y} \max_{x \geq 0} \left\{ c'x + y'(b - Ax) \right\}$$

$$\leq \min_{y} \left\{ y'b + \max_{x \geq 0} (c - A'y)' x \right\} = V_D.$$

It turns out that when either the primal or the dual problem is feasible, then the weak duality inequality is actually an equality. This fact, which follows quite directly from theorem B.1, is left as an exercise. □

Finally, we should also mention *complementary slackness* relations: if x and y are solutions to the primal and dual problem, respectively, and if one constraint of the dual program is not saturated, say $\sum_j A_{ij} y_j > c_i$ for some i, then we have $x_i = 0$.

B.3 LINK WITH ZERO-SUM GAMES

Consider a zero-sum game with two players. The sets $\mathcal{X}$ and $\mathcal{Y}$ of players 1 and 2's strategies are finite. If player 1 plays $x \in \mathcal{X}$ and player 2 plays $y \in \mathcal{Y}$, then player 1 receives transfer $\Phi_{xy} > 0$ from player 2. The value of this game to player 1 is

$$V = \max_{\sum_x p_x = 1} \min_{\sum_y q_y = 1} \sum_{x,y} p_x q_y \Phi_{xy}.$$

THEOREM B.3. *One has*

$$\frac{1}{V} = \min \left\{ \sum_x \alpha_x : \sum_x \alpha_x \Phi_{xy} \geq 1 \; \forall y \in \mathcal{Y} \right\}$$

$$= \max \left\{ \sum_y q_y : \sum_y q_y \Phi_{xy} \leq 1 \; \forall x \in \mathcal{X} \right\},$$

where these two linear programs are dual, and if (α_x) is a solution and (β_y) is the associated set of Lagrange multipliers, then

$$p_x^* = \alpha_x V, \quad q_y^* = \beta_y V$$

is an equilibrium in mixed strategy.

The proof of the theorem relies on the following very simple, but fundamental, lemma.

LEMMA B.4. *For $Z > 0$, the following holds:*

$$\max_{\lambda \geq 0} \min_{\mu \geq 0} \lambda + Z\left(\frac{\mu}{\lambda}\right) - \mu = Z.$$

PROOF. Indeed, this is equal to

$$\max_{\lambda \geq 0} \left\{ \lambda : \frac{Z}{\lambda} \geq 1 \right\} = Z.$$

$\square$

Let us now turn to the proof of the main result.

PROOF OF THEOREM B.3. Indeed,

$$V = \max_{\substack{\lambda \geq 0 \\ \sum_x p_x = 1}} \min_{\substack{\mu \geq 0 \\ \sum_y q_y = 1}} \lambda + \sum p_x q_y \Phi_{xy}\left(\frac{\mu}{\lambda}\right) - \mu$$

$$= \max_{\alpha \geq 0} \min_{\beta \geq 0} \frac{1}{\sum_x \alpha_x} + \sum_{xy} \alpha_x \beta_y \Phi_{xy} - \sum_y \beta_y$$

$$= \max_{\alpha \geq 0} \left\{ \frac{1}{\sum_x \alpha_x} : \sum_x \alpha_x \Phi_{xy} \geq 1 \, \forall y \right\}.$$

Hence if (α_x) is a solution and (β_y) is the associated set of Lagrange multipliers, then

$$p_x^* = \alpha_x V, \qquad q_y^* = \beta_y V$$

is an equilibrium in mixed strategy. Note that

$$\frac{1}{V} = \min \left\{ \sum_x \alpha_x : \sum_x \alpha_x \Phi_{xy} \geq 1 \, \forall y \right\}$$

$$= \max \left\{ \sum_y q_y : \sum_y q_y \Phi_{xy} \leq 1 \right\}$$

and these two problems are dual to one another. One has thus

$$\min \left\{ \sum_x p_x : \sum_x p_x \Phi_{xy} \geq V \right\} = 1 = \max \left\{ \sum_y q_y : \sum_y q_y \Phi_{xy} \leq V \right\}.$$

$\square$

B.4 REFERENCES AND NOTES

There are many introductions to linear programming; we personally like Vohra [150, chapter 4], which is concise and didactical. The reference for von Neumann's minimax theorem is [152]; for Sion's theorem, it is [140]. The classical reference for the linear programming formulation of zero-sum games is Dantzig [43]; see also von Neumann [153].

Appendix C

Quantiles and Copulas

Note that throughout this section, $\mathcal{U}$ will denote $\mathcal{U}\left([0,1]\right)$.

C.1 QUANTILES

Recall the definition of a *cumulative distribution function,*

$$F_P(x) = \Pr\left(X \leq x\right),$$

which is continuous if the distribution of X has no mass point. In general, F_P is nondecreasing and cadlag, that is, "continu à droite, limite à gauche"—which, translated from French, means that at every point,

- F_P is right continuous: $\lim_{x \to x_0^+} F_P(x) = F_P(x_0)$

and

- F_P has a limit from the left: $F_P(x^-) := \lim_{x \to x_0^-} F_P(x)$ exists.

The nice feature of nondecreasing cadlag functions is that they can be inversed in a generalized sense. Given F nondecreasing and cadlag, and for $t \in \mathrm{Im}\,(F)$, define

$$F^{-1}(t) = \inf\{x : F(x) > t\}, \tag{C.1}$$

which is also nondecreasing and cadlag on its domain. Then, F^{-1} is called the *generalized inverse* of F. The following lemma will be useful.

LEMMA C.1.   (i) *The inequalities*

$$F\left(F^{-1}(t)\right) \geq t \quad and \quad F^{-1}(F(x)) \geq x$$

hold in general, and hold with equality if and only if F is continuous at $F^{-1}(t)$ (respectively, F^{-1} is continuous at $F(x)$).
(ii) *If F^{-1} is continuous at t, then $F^{-1}(t) \leq x$ if and only if $t \leq F(x)$.*

PROOF. (i) The proof of (i) is obtained by noticing that

$$F(F^{-1}(t)) = \inf \{F(x) : F(x) > t\},$$

$$F^{-1}(F(x)) = \inf\{x' : F(x') > F(x)\},$$

where the fact that F is right continuous has been used in the first expression.

(ii) Assume $F^{-1}(t) \leq x$. Then $t \leq F\left(F^{-1}(t)\right) \leq F(x)$. Conversely, let us assume $t \leq F(x)$. Then either $F(x) = t$ or $F(x) > t$ holds. If $F(x) = t$, then F^{-1} is continuous at $F(x)$, and thus $F^{-1}(F(x)) = x$, thus $F^{-1}(t) = x$. If $F(x) > t$, then $x \geq F^{-1}(t)$ by the very definition (C.1). □

The quantile map of $X \sim P$ is classically defined as the generalized inverse of F_P, namely,

$$Q_P = F_P^{-1}.$$

It is easy to see that F_P is continuous if and only if P has no mass point, and F_P^{-1} is continuous if and only if P has no holes, that is, the probability of every nonempty interval is positive.

A basic result is the Skorokhod representation of random variables using quantile transforms.

LEMMA C.2 (Skorokhod representation). *If $X \sim P$, then there is $U \sim \mathcal{U}$ such that $X = F_P^{-1}(U)$ holds almost surely.*

In general $F_P(X)$ is not a uniform distribution, unless X has no mass point. For instance, if X is a Bernoulli distribution with parameter p, then $F_P(X)$ takes value 0 with probability $1 - p$ and 1 with probability p. This means that if X has mass points, one needs a larger probability space to represent U than to represent X. The following proof shows how to do this, based on the idea of Ferguson's "generalized distributional transform"[1] associated with distribution P, which is defined for $x \in \mathbb{R}$, $\lambda \in [0, 1]$ by

$$\tilde{F}_P(x, \lambda) = \Pr(X < x) + \lambda \Pr(X = x).$$

PROOF. Note that $\tilde{F}_P(x, \lambda) = t$ implies

$$\Pr(X < x) \leq t \leq \Pr(X \leq x). \tag{C.2}$$

Let $\tilde{U} \sim \mathcal{U}$ be independent from X, and define

$$U = \tilde{F}_P\left(X, \tilde{U}\right).$$

---

[1] See Ferguson [55], Rüschendorf [128].

It follows from (C.2) that

$$F_P(X^-) \leq U \leq F_P(X). \tag{C.3}$$

Letting

$$q = \Pr\left(X < F_P^{-1}(t)\right) \quad \text{and} \quad m = \Pr\left(X = F_P^{-1}(t)\right),$$

one has

$$1\{U \leq t\} = 1\left\{X < F_P^{-1}(t)\right\} + 1\left\{X = F_P^{-1}(t)\right\} 1\{q + \tilde{U}m \leq t\};$$

thus

$$\Pr(U \leq t) = q + m\Pr\left(\tilde{U} \leq \frac{t-q}{m}\right) = t,$$

hence $U \sim \mathcal{U}$. Finally, note that for any $t \in (F_P(x^-), F_P(x)]$, one has $F_P(x) = t$. As a result, $X = F_P^{-1}(U)$ holds almost surely. □

By contrast, the following lemma expresses that, if X has no mass point, then $F_P(X) \sim \mathcal{U}$.

LEMMA C.3. *If $X \sim P$ and P has no mass point, then $F_P(X) \sim \mathcal{U}$.*

PROOF. By lemma C.2, there is $U \sim \mathcal{U}$ such that $X = F_P^{-1}(U)$. Because P has no mass point, F_P is continuous. By lemma C.1(i), $F_P\left(F_P^{-1}(t)\right) = t$ for all t. Thus $F_P\left(F_P^{-1}(U)\right) \sim \mathcal{U}$. □

C.2 COPULAS

We now wonder whether we can provide *jointly* a representation of X as $F_P^{-1}(U)$, and of Y as $F_Q^{-1}(V)$, namely, whether there exists a pair (U, V) of uniform random variables such that $\left(F_P^{-1}(U), F_Q^{-1}(V)\right)$ has the same distribution as (X, Y).

The first building block is Sklar's lemma. It requires the definition of a copula.

DEFINITION C.4. *A copula is the c.d.f. of a joint distribution with uniform marginals. We shall denote by*

$$\mathcal{C} := \{F_\pi : \pi \in M(\mathcal{U}, \mathcal{U})\}$$

the set of copulas.

LEMMA C.5 (Sklar). *If $\pi \in \mathcal{M}(P, Q)$, then there exists a copula $C \in \mathcal{C}$ such that the c.d.f. of π can be written as*

$$F_\pi(x, y) = C\left(F_P(x), F_Q(y)\right).$$

The interpretation of this result is clear: one can "factor" the information on the joint distribution of (X, Y) into

- the information on the marginal distributions of X and Y, which is captured by F_P and F_Q; and
- the information on the joint dependence between X and Y, which is captured by C.

Copulas are very useful in practice because they allow us to obtain semi-parametric models of the distribution of a random vector, where the marginal distributions P and Q are left nonparametric, and the joint distribution is captured by a copula belonging to a parametric distribution. For instance, one of the most commonly used copulas is the Gaussian copula of correlation ρ, which is the copula associated with the joint distribution of (X, Y), a bivariate Gaussian vector with correlation ρ.

There are many proofs of lemma C.5. A number of proofs establish the result when X and Y have no atom, which is immediate, and then use an approximation argument. Our proof is based on the generalized distributional transform introduced in the proof of lemma C.2.

PROOF. Consider $(X, Y) \sim \pi \in \mathcal{M}(P, Q)$. Let $\tilde{U}, \tilde{V}$ be a pair of i.i.d. random variables independent from (X, Y). Introduce

$$U = F_X\left(X, \tilde{U}\right) \quad \text{and} \quad V = F\left(Y, \tilde{V}\right),$$

and define C as the c.d.f. of the joint distribution of (U, V). Then, the argument in the proof of lemma C.2 can be adapted to show that $X = F_P^{-1}(U)$ and $Y = F_Q^{-1}(V)$. Next,

$$F_\pi(x, y) = \Pr\left(X \leq x, Y \leq y\right) = \mathbb{E}\left[\mathbf{1}\left\{F_P^{-1}(U) \leq x\right\} \mathbf{1}\left\{F_Q^{-1}(V) \leq y\right\}\right].$$

But almost surely, F_P^{-1} and F_Q^{-1} are continuous at U and V respectively; thus by lemma C.1(ii),

$$F_\pi(x, y) = \mathbb{E}\left[\mathbf{1}\left\{U \leq F_P(x)\right\} \mathbf{1}\left\{V \leq F_Q(y)\right\}\right] = C\left(F_P(x), F_Q(y)\right).$$

$\square$

As soon as either P or Q has atoms, the copula C associated with $\pi \in \mathcal{M}(P, Q)$ is not unique. Hence it would be more correct to refer to "a" copula associated with π rather than "the" copula. However, we shall assume that we break indeterminacy and we will associate a unique copula to any $\pi \in \mathcal{M}(P, Q)$.

As an immediate consequence of lemma C.5, we can obtain a bivariate analogue of lemma C.2.

LEMMA C.6. *Let* $\pi \in \mathcal{M}(P, Q)$, *and consider* C_π, *the corresponding copula. Let* (U, V) *be a random pair of variables with joint c.d.f.* C_π. *Then*

$$\left(F_P^{-1}(U), F_Q^{-1}(V)\right) \sim \pi.$$

In other words, every element of $\mathcal{M}(P, Q)$ can be interpreted as the distribution of a random pair of the form $\left(F_P^{-1}(U), F_Q^{-1}(V)\right)$, where U and V are uniform random variables.

C.3 REFERENCES AND NOTES

Many of the notions recalled here can be found in any intermediate-level texts on probability and statistics. For notions related to couplings, the reader might refer to Rachev and Rüschendorf [118]. For more results and examples on copulas, see Nelsen [112].

═══ Appendix D ═══

Basics of Convex Analysis

D.1 CONVEX SETS

The space here is $\mathbb{R}^d$.

DEFINITION D.1 (Convex set). *A set C is convex if for all $x, y \in C$ and $t \in [0, 1]$, $tx + (1 - t)y \in C$.*

THEOREM D.2 (Separating hyperplane theorem). *Let C be a nonempty closed convex set and $z \notin C$. Then there is a hyperplane strictly separating C and z.*

In what follows, X is a set of $\mathbb{R}^d$.

DEFINITION D.3 (Convex hull). *The convex hull of X, denoted $\mathrm{conv}(X)$, is the set of $x = \sum_{i=1}^{p} \alpha_i x_i$ for all $x_1, \ldots, x_p$ in X and $\alpha_i \geq 0$ such that $\sum_{i=1}^{p} \alpha_i = 1$.*

The convex hull of X is the smallest convex set containing X. When X is finite, $\mathrm{conv}(X)$ is always closed. When it is infinite, it may be open (for instance, the open ball).

Hence, next we define the closed convex hull of X as the closure of $\mathrm{conv}(X)$.

DEFINITION D.4 (Closed convex hull). *The closed convex hull of X, denoted $\mathrm{cch}(X)$, is the closure of the convex hull of X.*

In what follows, C is a closed convex set of $\mathbb{R}^d$.

DEFINITION D.5 (Extreme point). *An extreme point of C is a point $z \in C$ such that $z = tx + (1 - t)y \in C$, $t \in (0, 1)$ implies $x = y = z$.*

DEFINITION D.6 (Boundary point). *A boundary point of C is a point x such that there exists a vector b such that*

$$x'b = \max_{y \in C} y'b.$$

The set of boundary points of X is also the topological boundary of C.

Extreme points are boundary points. The converse is false in general. However, the set of extreme points of C is the set of points $x \in C$ such that there exists a vector b such that x is the unique maximizer of $\max_{y \in C} y \cdot b$.

THEOREM D.7 (Krein–Milman). *A compact convex set is the convex hull of its extreme points.*

D.2 CONVEX FUNCTIONS

Our definition of convex functions slightly differs from the commonly used one, as we shall add a requirement of nonempty domain.

DEFINITION D.8 (Convex function). *A function $\varphi : \mathbb{R}^d \to \mathbb{R} \cup \{+\infty\}$ is called convex when for all $x, y \in \mathbb{R}^d$ and $t \in [0, 1]$, $\varphi(tx + (1 - t)y) \leq t\varphi(x) + (1 - t)\varphi(y)$ and it is not identically $+\infty$.*

The domain of φ is the set of x such that $\varphi(x) < +\infty$. It is a convex set. If $(y_1, v_1, \ldots, y_p, v_p)$ are vectors in $\mathbb{R}^d$, then

$$\varphi(x) = \max_k \left(x' y_k - v_k \right)$$

is a piecewise-affine convex function.

To a given set X, one can associate a function ι_X such that

$$\iota_X(x) = \begin{cases} 0 & \text{if } x \in X, \\ +\infty & \text{otherwise.} \end{cases}$$

Note that ι_X is a convex function if and only if X is a convex set.

D.2.1 Differentiability

A convex function is continuous and locally Lipschitz on the interior of its domain, hence the set of points where it is not differentiable is of Lebesgue measure zero. Let $\nabla \varphi(x)$ be the gradient of φ at x; it is the vector of derivatives $(\partial \varphi(x)/\partial x_1, \ldots, \partial \varphi(x)/\partial x_d)$.

For all points x where φ is differentiable, one has

$$\varphi(y) \geq \varphi(x) + \nabla \varphi(x)'(y - x).$$

This motivates the following definition of the subdifferential, as a generalization of this idea.

DEFINITION D.9 (Subdifferential). *Let $\varphi : \mathbb{R}^d \to \mathbb{R}$. We define the subdifferential of φ at x, denoted $\partial \varphi(x)$, as the set of $y \in \mathbb{R}^d$ such that for all $\tilde{x} \in \mathbb{R}^d$, $\varphi(\tilde{x}) \geq \varphi(x) + y'(\tilde{x} - x)$.*

As we shall see, if φ is differentiable at x, then $\partial \varphi(x) = \{\nabla \varphi(x)\}$.

D.2.2 Conjugate functions

We start with the definition of Legendre–Fenchel transforms.

DEFINITION D.10 (Legendre–Fenchel transform). *For a function φ that is not identically $+\infty$, one defines its Legendre–Fenchel transform (or convex conjugate function) φ^* as*

$$\varphi^*(y) = \sup_{x \in \mathbb{R}^d} \left(x'y - \varphi(x) \right).$$

PROPOSITION D.11 (Legendre transform). *The following holds:*

 (i) *φ^* is convex.*
 (ii) *$\varphi_1 \le \varphi_2$ implies $\varphi_1^* \ge \varphi_2^*$.*
 (iii) *Fenchel's inequality: $\varphi(x) + \varphi^*(y) \ge x'y$.*

Let us take some examples.

Example D.1 (Quadratic functions). (i) *For $\varphi(x) = \frac{1}{2}|x|^2$, one gets $\varphi^*(y) = \frac{1}{2}|y|^2$.*
 (ii) *For $\varphi(x) = \frac{1}{2}\sum_i \lambda_i x_i^2$, $\lambda_i > 0$, one gets $\varphi^*(y) = \frac{1}{2}\sum_i \lambda_i^{-1} y_i^2$.*
 (iii) *More generally, for $\varphi(x) = \frac{1}{2}x'\Sigma x$, where Σ is a positive-definite matrix, one has*

$$\varphi^*(y) = \frac{y'\Sigma^{-1}y}{2}.$$

Example D.2 (Entropy). *Consider*

$$\varphi(x) = \begin{cases} \sum_i x_i \ln x_i & \text{for } x \ge 0,\ \sum_i x_i = 1, \\ +\infty & \text{otherwise.} \end{cases}$$

Then

$$\varphi^*(y) = \ln \left(\sum_i e^{y_i} \right).$$

Example D.3 (Power of the norm). *Let $p > 1$ and*

$$\varphi(x) = \frac{1}{p} \|x\|^p,$$

where $\|\cdot\|$ is the Euclidean norm. Then

$$\varphi^*(y) = \frac{1}{q} \|y\|^q,$$

where $q > 1$ such that $\frac{1}{p} + \frac{1}{q} = 1$.

For what follows, note that for $r > 0$, the function

$$\varphi(x) = \left(\sum_i x_i^r \right)^{1/r}$$

is concave if $r \leq 1$ and convex if $r \geq 1$.

Example D.4 (CES function). *For $x \in \mathbb{R}_+^d$, define*

$$\varphi(x) = \left(\sum_{i=1}^d x_i^{(\alpha-1)/\alpha} \right)^{\alpha/(\alpha-1)},$$

which is called a constant elasticity of substitution (CES) function with elasticity parameter α. Letting $r = (\alpha - 1)/\alpha < 1$, it turns out that $\varphi(x)$ is convex for $r \geq 1$ and concave for $r \leq 1$. Then, when $r > 1$, φ can be represented as

$$\varphi(x) = \sup_{y \geq 0} \left\{ \sum_{i=1}^d x_i y_i : \sum_{i=1}^d y_i^{1-\alpha} = 1 \right\}.$$

Hence, when $r > 1$, $\varphi^*(y) = 0$ if $\sum_{i=1}^d y_i^{1-\alpha} \leq 1$, $+\infty$ otherwise. A similar logic shows that when $r \leq 1$,

$$\varphi(x) = \inf_{y \geq 0} \left\{ \sum_{i=1}^d x_i y_i : \sum_{i=1}^d y_i^{1-\alpha} = 1 \right\}.$$

D.2.3 Duality

We have in general

$$x'y \leq \varphi(x) + \varphi^*(y)$$

and equality holds if and only if $y \in \partial\varphi(x)$, or equivalently $x \in \partial\varphi(y)$. In particular, if φ is differentiable at x, if $y = \nabla\varphi(x)$, and if φ^* is differentiable at y, one gets

$$x = \nabla\varphi^*(y).$$

If φ is convex and not identically $+\infty$, then one gets

$$\varphi^{**} = \varphi.$$

PROPOSITION D.12 (Convex envelope). *The following holds:*

(i) $\varphi^{**} \leq \varphi$, with equality if and only if φ is convex.
(ii) φ^{**} is the greatest minorant of φ, that is, $\varphi^{**}(x) = \sup\{g(x) : g \leq \varphi \text{ and } g \text{ is convex}\}$.

PROPOSITION D.13. *Let $\varphi : \mathbb{R}^d \to \mathbb{R}$ be a convex function. The following statements are equivalent:*

    (i) $\varphi(x) + \varphi^*(y) = x'y$.
    (ii) $y \in \partial u(x)$.
    (iii) $x \in \partial u^*(y)$.

PROPOSITION D.14. *Let φ be a convex function. Then the following statements are equivalent:*

    (i) φ *is differentiable.*
    (ii) *For each $x \in \mathbb{R}^d$, $\partial\varphi(x)$ contains a single element.*
    (iii) $\varphi \in C^1$.
    (iv) φ^* *is strictly convex.*

D.2.4 Support functions

The *support function S_Y of Y* is defined as

$$S_Y(x) = \sup_{y \in Y} x'y$$

for any x in Y. It is a convex function, and it is homogeneous of degree 1. Note that

$$S_Y = \iota_Y^*.$$

Moreover, $S_Y = S_{\text{cch}(Y)}$ where cch (Y) is the closed convex hull of Y, and $\partial S_Y(0) = \text{cch}(Y)$.

D.2.5 Almost Everywhere Differentiability

The following two results are extremely useful. They state that the first and second derivatives of convex functions exist almost everywhere.

PROPOSITION D.15 (Rademacher's theorem). *Let $\varphi : \mathbb{R}^d \to \mathbb{R}$ be a convex function. Then $\nabla\varphi(x)$ exists Lebesgue-almost everywhere.*

PROPOSITION D.16 (Alexandrov's theorem). *Let $\varphi : \mathbb{R}^d \to \mathbb{R}$ be a convex function. Then $D^2\varphi(x)$ exists Lebesgue-almost everywhere.*

Actually, the theorems that bear these names are significantly stronger and more general. The statements we provide here are corollaries which suffice for our purposes.

D.3 REFERENCES AND NOTES

The treatise by Rockafellar [122] and the one by Ekeland and Temam [53] are classics. Hiriart-Urrut and Lemaréchal [79] is a gentle introduction. The present exposition borrows from the short and efficient reminder in [148, pp. 50–52].

═══ Appendix E ═══

McFadden's Generalized Extreme Value Theory

Let **F** be a cumulative distribution such that the function g defined by

$$g(x_1, \ldots, x_n) = -\log \mathbf{F}(-\log x_1, \ldots, -\log x_n) \qquad \text{(E.1)}$$

is positive homogeneous of degree 1. (This inverts into $\mathbf{F}(u_1, \ldots, u_n) = \exp(-g(e^{-u_1}, \ldots, e^{-u_n}))$.) By a theorem of McFadden [104, 105], we have the following result.

THEOREM E.1 (McFadden). *Let $(\varepsilon_i)_{1 \leq i \leq n}$ be a random vector with c.d.f. **F**, and define*

$$Z - \max_{i=1,\ldots,n} \{u_i + \varepsilon_i\}.$$

Then Z is a $(\log g(e^{u_1}, \ldots, e^{u_n}), 1)$ -Gumbel distribution. In particular,

$$\mathbb{E}\left[\max_{i=1,\ldots,n} \{u_i + \varepsilon_i\}\right] = \log g\left(e^{u_1}, \ldots, e^{u_n}\right) + \gamma,$$

where γ is the Euler constant, $\gamma \simeq 0.5772$.

PROOF. Let G be the c.d.f. of $Z = \max_{i=1,\ldots,n} \{u_i + \varepsilon_i\}$, and let $\varphi(z) := \exp(-e^{-z})$ be the c.d.f. of the $(0,1)$-Gumbel distribution. One has

$$G(z) = \Pr\left(\max_{i=1,\ldots,n} \{u_i + \varepsilon_i\} \leq z\right) = \Pr\left(\forall i: \varepsilon_i \leq z - u_i\right)$$

$$= \mathbf{F}(z - u_1, \ldots, z - u_n) = \exp\left(-g\left(e^{u_1-z}, \ldots, e^{u_n-z}\right)\right)$$

$$= \exp\left(-e^{-z}g\left(e^{u_1}, \ldots, e^{u_n}\right)\right) = \varphi\left(z - \log g\left(e^{u_1}, \ldots, e^{u_n}\right)\right).$$

$\square$

E.1 REFERENCES AND NOTES

A good textbook reference for this topic is [28].

References

[1] Ahuja, R., Magnanti, T., and Orlin, J. (1993). *Network Flows: Theory, Algorithms and Applications*. Prentice-Hall.

[2] Alkan, A. and Gale, D. (1990). "The core of the matching game." *Games and Economic Behavior*, 2(3):203–212.

[3] Anderson, S., de Palma, A., and Thisse, J.-F. (1992). *Discrete Choice Theory of Product Differentiation*. MIT Press.

[4] Armstrong, M. (1996). "Multiproduct nonlinear pricing." *Econometrica* 64:51–75.

[5] Armstrong, M. and Rochet, J.-C. (1999). "Multi-dimensional screening: A user's guide." *European Economic Review* 43(4-6):959–979.

[6] Artstein, Z. (1983). "Distributions of random sets and random selections." *Israel Journal of Mathematics* 46:313–324.

[7] Aurenhammer, F. (1987). "Power diagrams: Properties, algorithms and applications." *SIAM Journal on Computing* 16:78–96.

[8] Aurenhammer, F. (1991). "Voronoi diagrams – A survey of a fundamental geometric data structure." *ACM Computing Surveys* 23:345–405.

[9] Aurenhammer, F., Hoffmann, F., and Aronov, B. (1988). "Minkowski-type theorems and least-squares clustering." *Algorithmica* 20:61–76.

[10] Bates, D. and Maechler, M. (2015). "`Matrix`: Sparse and dense matrix classes and methods. " R package version 1.2-3. `https://CRAN.R-project.org/package=Matrix`.

[11] Becker, G. (1973). "A theory of marriage, part I." *Journal of Political Economy*, 81:813–846.

[12] Becker, G. (1993). *A Treatise of the Family*. Harvard University Press.

[13] Beckmann, M. J. (1952). "A continuous model of transportation." *Econometrica* 20:642–660.

[14] Benamou, J. D., Carlier, G., Cuturi, M., Nenna, N., and Peyré, G. (2015). "Iterative Bregman projections for regularized transportation problems." *SIAM Journal on Scientific Computing* 37(2):A1111–A1138.

[15] Beresteanu, A., Molchanov I., and Molinari, F. (2011). "Sharp identification regions in models with convex moment predictions." *Econometrica* 79(6): 1785–1821.

[16] Bertsekas, D. (1979). "A distributed algorithm for the assignment problem." MIT working paper.

[17] Bertsekas, D. (1981). "A new algorithm for the assignment problem." *Mathematical Programming* 21:152–171.

[18] Bertsekas, D. (1998). *Network Optimization: Continuous and Discrete Models*. Athena Scientific.

[19] Birkhoff, G. (1946). "Three observations on linear algebra." *Universidad Nacional de Tucumán. Revista. Serie A* 5:147–151.

[20] Bonnet, O., Galichon, A., and Shum, M. (2015). "Yoghurt chooses consumers: Identification of random utility models via two-sided matching." Preprint.

[21] Borchers, H. W. (2015). "adagio: Discrete and global optimization routines." R package version 0.6.3. https://CRAN.R-project.org/package=adagio.

[22] Bosc, D. (2012). "Three essays on modeling the dependence between financial assets." PhD thesis, École Polytechnique.

[23] Breeden, D. and Litzenberger R. (1978). "Prices of state-contingent claims implicit in option prices." *Journal of Business* 51(4):621–651.

[24] Brenier, Y. (1987). "Décomposition polaire et réarrangement monotone des champs de vecteurs." *C.R. Acad. Sci. Paris, Série I* 305:805–808.

[25] Brenier, Y. (1991). "Polar factorization and monotone rearrangement of vector-valued functions." *Communications on Pure and Applied Mathematics* 44: 375–417.

[26] Burchard, A. and Hajaiej, H. (2006). "Rearrangement inequalities for functionals with monotone integrands." *Journal of Functional Analysis* 233:561–582.

[27] Burkard, R., Dell'Amico, M., and Martello, S. (2012). *Assignment Problems.* SIAM.

[28] Cameron, A. Colin and Trivedi, P. (2005). *Microeconometrics: Methods and Applications.* Cambridge University Press.

[29] Carlier, G. (2001). "A general existence result for the principal–agent problem with adverse selection." *Journal of Mathematical Economics* 35:129–150.

[30] Carlier, G. (2002). "Nonparametric adverse selection problems." *Annals of Operations Research* 114:71–82.

[31] Carlier, G. (2003). "Duality and existence for a class of mass transportation problems and economic applications." *Advances In Mathematical Economics* 5:1–21.

[32] Carlier, G. (2010). Lecture notes on "Optimal transportation and economic applications."

[33] Carlier, G., Chernozhukov, V., and Galichon, A. "Vector quantile regression: An optimal transport approach." *Annals of Statistics*, forthcoming.

[34] Carlier, G., Galichon, A., and Santambrogio, F. (2010). "From Knothe's transport to Brenier's map." *SIAM Journal on Mathematical Analysis* 41(6):2554–2576.

[35] Chernozhukov, V., Fernandez-Val, I., and Galichon, A. (2009). "Improving estimates of monotone functions by rearrangement." *Biometrika* 96:559–575.

[36] Chernozhukov, V., Fernandez-Val, I., and Galichon, A. (2010). "Quantile and probability curves without crossing." *Econometrica*, 78(3):1093–1125.

[37] Chernozhukov, V., Galichon, A., Hallin, M., and Henry, M. (2015). "Monge–Kantorovich depth, quantiles, ranks and signs." *Annals of Statistics*, forthcoming.

[38] Chernozhukov, V., Galichon, A., Henry, M., and Pass, B. (2015). "Single market nonparametric identification of multi-attribute hedonic equilibrium models." Preprint. Available from http://ssrn.com/abstract=2392549.

[39] Chiappori, P.-A., McCann, R., and Nesheim, L. (2010). "Hedonic price equilibria, stable matching, and optimal transport: Equivalence, topology, and uniqueness." *Economic Theory* 42(2):317–354.

[40] Chiong, K., Galichon, A., and Shum, M. (2015). "Duality in dynamic discrete choice models." *Quantitative Economics*, forthcoming.

[41] Crawford, V. P. and Knoer, E. M. (1981). "Job matching with heterogeneous firms and workers." *Econometrica*, 49(2):437–50.

[42] Cuesta-Albertos, J. Ruschendorf, L., and Tuero-Diaz, A. (1993). "Optimal coupling of multivariate distribution and stochastic processes." *Journal of Multivariate Analysis* 46:335–361.

[43] Dantzig, G. (1951). "A proof of the equivalence of the programming problem and the game problem." In *Activity Analysis of Production and Allocation*, T. C. Koopmans (ed), pp. 330–335. Wiley.

[44] Dantzig, G. (1963). *Linear Programming and Extensions*. Princeton University Press.

[45] Deming, W. and Stephan, F. (1940). "On a least squares adjustment of a sampled frequency table when the expected marginal totals are known." *Annals of Mathematical Statistics* 11(4):427–444.

[46] Dowson, D. C. and Landau, B. V. (1982). "The Fréchet distance between multivariate normal distributions." *Journal of Multivariate Analalysis* 12:450–455.

[47] Dupuy, A. and Galichon, A. (2014). "Personality traits and the marriage market." *Journal of Political Economy* 122(6):1271–1319.

[48] Dupuy, A., Galichon, A., and Henry, M. (2014). "Entropy methods for identifying hedonic models." *Mathematics and Financial Economics* 8(4):405–416.

[49] Ekeland, I. (2010). "Existence, uniqueness and efficiency of equilibrium in hedonic markets with multidimensional types." *Economic Theory* 42:275–315.

[50] Ekeland, I. (2010). "Notes on optimal transportation." *Economic Theory* 42(2):437–459.

[51] Ekeland, I. Galichon, A., and Henry, M. (2012). "Comonotonic measures of multivariate risks." *Mathematical Finance* 22(1):109–132.

[52] Ekeland, I., Heckman, J., and Nesheim, L. (2004). "Identification and estimation of hedonic models." *Journal of Political Economy* 112(S1):S60–S109.

[53] Ekeland, I. and Temam, R. (1976). *Convex Analysis and Variational Problems*. North-Holland, Amsterdam-Oxford.

[54] Feenstra, R. and Levinsohn, J. (1995). "Estimating markups and market conduct with multidimensional product attributes." *Review of Economic Studies* 62:19–52.

[55] Ferguson, T. S. (1967). *Mathematical Statistics: A Decision Theoretic Approach*. New York: Academic Press.

[56] Figalli, A., Kim, Y.-H., and McCann, R. (2011). "When is multidimensional screening a convex program?" *Journal of Economic Theory* 146(2):454–478.

[57] Fryer, R. and Holden, R. (2011). "Measuring the compactness of political districting plans." *Journal of Law and Economics* 54(3):493–535.

[58] Gabaix, X. and Landier, A. (2008). "Why has CEO pay increased so much?" *Quarterly Journal of Economics* 123(1):49–100.

[59] Gale, D. (1984). "Equilibrium in a discrete exchange economy with money." *International Journal of Game Theory*, 13(1):61–64.

[60] Gale, D. and Shapley, L. (1962). "College admissions and the stability of marriage." *American Mathematical Monthly* 69(1):9–15.

[61] Galichon, A. (2011). "Theoretical and empirical aspects of matching markets." Unpublished lecture notes, Columbia University.

[62] Galichon, A. (2015). Online material for *Optimal Transport Methods in Economics.* Available from `http://press.princeton.edu/titles/10870.html`.

[63] Galichon, A. and Henry, M. (2006). "Inference in incomplete models." Unpublished manuscript. Available from `http://ssrn.com/abstract=886907`.

[64] Galichon, A. and Henry, M. (2011). "Set identification in models with multiple equilibria." *Review of Economic Studies* 78(4):1264–1298.

[65] Galichon, A. and Henry, M. (2012). "Dual theory of choice with multivariate risks." *Journal of Economic Theory* 147(4):1501–1516.

[66] Galichon, A., Henry-Labordère, P., and Touzi, N. (2014). "A stochastic control approach to no-arbitrage bounds given marginals, with an application to lookback options." *Annals of Applied Probability* 24(1):312–336.

[67] Galichon, A., Kominers, S., and Weber, S. (2015). "Costly concessions. An empirical framework for matching with imperfectly transferable utility." Working paper.

[68] Galichon, A. and Salanié, B. (2010). "Matching with trade-offs: Revealed preferences over competing characteristics." Unpublished manuscript. Available from `http://ssrn.com/abstract=1487307`.

[69] Galichon, A. and Salanié, B. (2014). "Cupid's invisible hand: Social surplus and identification in matching models." Preprint. Available from `http://ssrn.com/abstract=1804623`.

[70] Galichon, A. and Weber, S. (2015). "`TraME`: Transportation methods for econometrics." R package. `https://github.com/TraME-Project/TraME`.

[71] Gangbo, W. and McCann, R. (1996). "The geometry of optimal transportation." *Acta Mathematica* 177(2):113–161.

[72] Grady, L. and Polimeni, J. (2010). *Discrete Calculus: Applied Analysis on Graphs for Computational Science.* Springer.

[73] Gretsky, N., Ostroy, J., and Zame, W. (1992). "The nonatomic assignment model." *Economic Theory* 2(1):103–127.

[74] Gurobi optimizer. `www.gurobi.com`.

[75] Habel, K., Grasman, R., Gramacy, R., Stahel, A., and Sterratt, D. (2015). "`geometry`: Mesh generation and surface tesselation." R package version 0.3-6. `https://CRAN.R-project.org/package=geometry`.

[76] Hardy, G., Littlewood, J., and Pólya, G. (1952). *Inequalities.* Cambridge University Press.

[77] Hatfield, J. W. and Milgrom, P. R. (2005). "Matching with contracts." *American Economic Review*, 95(4):913–935.

[78] Heckman, J. J., Matzkin, R. L., and Nesheim, L. (2010). "Nonparametric identification and estimation of nonadditive hedonic models." *Econometrica*, 78(5):1569–1591.

[79] Hiriart-Urrut, J.-B. and Lemaréchal, C. (2001). *Fundamentals of Convex Analysis.* Springer.

[80] Hornik, K. (2015). "`clue`: Cluster ensembles." R package version 0.3-50. `https://CRAN.R-project.org/package=clue`.

[81] Kaneko, M. (1982). "The central assignment game and the assignment markets." *Journal of Mathematical Economics*, 10(2-3):205–232.

[82] Kantorovich, L. (1939). "Mathematical methods in the organization and planning of production." Leningrad University. Reprint in *Management Science* 6:366–422 (1959–60).

[83] Kantorovich, L. (1942). "On the translocation of masses." *Dokl. Akad. Nauk SSSR* 37(7–8):227–229. Reprinted in *Journal of Mathematical Sciences* 133(4): 1381–1382 (2006).

[84] Kantorovich, L. (1948). "On a problem of Monge." *Uspekhi Mat. Nauk* 3:225–226 (in Russian). English translation in *Journal of Mathematical Sciences* 133(4):1383 (2006).

[85] Kantorovich, L. and Rubinstein, G. (1958). "On a space of completely additive functions." *Vestn. Leningrad Univ.* 13(7):52–59.

[86] Kelso Jr., A. S. and Crawford, V. P. (1982). "Job matching, coalition formation, and gross substitutes." *Econometrica*, 50(6):1483–1504.

[87] Knott, M. and Smith, C. S. (1984). "On the optimal mapping of distributions." *Journal of Optimization Theory and its Applications* 43:39–49.

[88] Koenker, R. (2005). *Quantile Regression.* Cambridge University Press.

[89] Koopmans, T. (1947). "Optimum utilization of the transportation system." *Proceedings of the International Statistics Conference* 5:136–146.

[90] Koopmans, T. and Beckmann, M. (1957). "Assignment problems and the location of economic activities." *Econometrica* 25(1):53–76.

[91] Laffont, J.-J. (1988). *Fundamentals of Public Economics.* MIT Press.

[92] Lancaster, K. J. (1966). "A new approach to consumer theory." *Journal of Political Economy* 74(2):132–157.

[93] Landsberger, M. and Meilijson, I. I. (1994). "Comonotone allocations, Bickel Lehmann dispersion and the Arrow–Pratt measure of risk aversion." *Annals of Operation Research* 52:97–106.

[94] Léonard, C. (2014). "A survey of the Schrödinger problem and some of its connections with optimal transport." *Discrete Contin. Dyn. Syst. A* 34(4): 1533–1574.

[95] Lloyd, S. (1982). "Least squares quantization in PCM." *IEEE Trans. Inform. Theory* 28:129–137.

[96] Lorentz, G. (1953). "An inequality for rearrangements." *American Mathematical Monthly* 60:176–179.

[97] Magnus, J. and Neudecker, H. (1988). *Matrix Differential Calculus with Applications in Statistics and Econometrics.* Wiley.

[98] Matheron, G. (1975). *Random Sets and Integral Geometry.* Wiley.

[99] Matzkin, R. (2003). "Nonparametric estimation of nonadditive random functions." *Econometrica* 71:1339–1375.

[100] Matzkin, R. (2013). "Nonparametric identification in structural econometric models." *Annual Review of Economics* 5:457–486.

[101] McAfee, R. P. and McMillan, J. (1988). "Multidimensional incentive compatibility and mechanism design." *Journal of Economic Theory* 46(2):335–354.

[102] McCann, R. (1995). "Existence and uniqueness of monotone measure-preserving maps." *Duke Mathematical Journal* 80(2):309–323.

[103] McCann, R. and Guillen, N. (2011). "Five lectures on optimal transportation: Geometry, regularity and applications." In *Analysis and Geometry of Metric Measure Spaces. Lecture Notes of the 50th Séminaire de Mathématiques Supérieures*

(SMS), Montréal, Canada, June 27 – July 8, 2011, pp. 145–180 (2013). Providence, RI: American Mathematical Society (AMS).

[104] McFadden, D. (1976). "The mathematical theory of demand models." In *Behavioral Travel-Demand Models*, P. Stopher and A. Meyburg (eds.), pp. 305–314. D. C. Heath.

[105] McFadden, D. (1978). "Modelling the choice of residential location." In *Spatial Interaction Theory and Residential Location*, A. Karlquist et al. (ed.), pp. 75–96. Amsterdam: North-Holland.

[106] Mérigot, Q. (2011). "A multiscale approach to optimal transport." *Computer Graphics Forum* 30(5):1583–1592.

[107] Mérigot, Q. and Oudet, E. (2014). "Handling convexity-like constraints in variational problems." *SIAM Journal on Numerical Analysis* 52(5):2466–2487.

[108] Molchanov, I. (2005). *Theory of Random Sets*. Springer.

[109] Monge, G. (1781). "Mémoire sur la théorie des déblais et des remblais." In *Histoire de l'Académie Royale des Sciences de Paris*, pp. 666–704.

[110] Müller, A. and Stoyan, D. (2002). *Comparison Methods for Stochastic Models and Risks*. Wiley.

[111] Mussa, M. and Rosen, S. (1978). "Monopoly and product quality." *Journal of Economic Theory* 18:301–317.

[112] Nelsen, R. (2006). *An Introduction to Copulas*. Springer.

[113] Noldeke, G. and Samuelson, L. (2015). "The implementation duality." Working paper.

[114] Ok, E. *Elements of Order Theory*. https://sites.google.com/a/nyu.edu/efeok/books.

[115] Olkin, I. and Pukelsheim, F. (1982). "The distance between two random vectors with given dispersion matrices." *Linear Algebra and its Applications* 48:257–263.

[116] Queyranne, M. (2011). Personal communication.

[117] R-core team (2015). "parallel." R-core package since R version 2.14.0. https://www.r-project.org/.

[118] Rachev, S. T. and Rüschendorf, L. (1998). *Mass Transportation Problems. Vol. I: Theory. Vol. II: Applications*. Springer.

[119] Rochet, J.-C. (1987). "A necessary and sufficient condition for rationalizability in a quasi-linear context." *Journal of Mathematical Economics* 16:191–200.

[120] Rochet, J.-C. and Choné, P. (1998). "Ironing, sweeping, and multidimensional screening." *Econometrica* 66(4):783–826.

[121] Rochet, J.-C. and Stole, L. (2003). "The economics of multidimensional screening." In *Advances in Economics and Econometrics*, M. Dewatripont, L.P. Hansen, S.J. Turnovsky (eds.), pp. 115–150. Cambridge University Press.

[122] Rockafellar, R. T. (1970). *Convex Analysis*. Princeton University Press.

[123] Rockafellar, R. T. (1984). *Network Flows and Monotropic Optimization*. Wiley.

[124] Rosen, S. (1974). "Hedonic prices and implicit markets: Product differentiation in pure competition." *Journal of Political Economy* 82(1):34–55.

[125] Roth, A. and Sotomayor, M. (1990). *Two-Sided Matching. A Study in Game-Theoretic Modeling and Analysis*. Cambridge University Press.

[126] Rothschild, M. and Stiglitz, J. (1970). "Increasing risk: I. A definition." *Journal of Economic Theory* 2:225–243.

[127] Rüschendorf, L. (1995). "Convergence of the iterative proportional fitting procedure." *Annals of Statistics* 23:1160–1174.

[128] Rüschendorf, L. (2009). "On the distributional transform, Sklar's theorem, and the empirical copula process." *Journal of Statistical Planning and Inference* 139(11):3921–3927.

[129] Rüschendorf, L. and Rachev, S. (1990). "A characterization of random variables with minimum L^2-distance." *Journal of Multivariate Analysis* 32:48–54.

[130] Rüschendorf, L. and Thomsen, W. (1993). "Note on the Schrödinger equation and I-projections." *Statistics and Probability Letters* 17:369–375.

[131] Ryff, J. (1970). "Measure preserving transformations and rearrangements." *Journal of Mathematical Analysis Applied* 31:449–458.

[132] Santambrogio, F. (2015). *Optimal Transport for Applied Mathematicians. Calculus of Variations, PDEs, and Modeling.* Birkhaüser.

[133] Sattinger, M. (1975). "Comparative advantage and the distributions of earnings and abilities." *Econometrica,* 43(3):455-468.

[134] Sattinger, M. (1979). "Differential rents and the distribution of earnings." *Oxford Economic Papers* 31(1):60–71.

[135] Sattinger, M. (1993). "Assignment models of the distribution of earnings." *Journal of Economic Literature* 31(2):831–880.

[136] Schuhmacher, D., with substantial contributions of code by B. Baehre and C. Gottschlich (2015). "`transport`: Optimal transport in various forms." R package version 0.7-0. `https://CRAN.R-project.org/package=transport`.

[137] Shaked, M. and Shanthikumar, J. G. (2006). *Stochastic Orders.* Springer.

[138] Shapley, L. and Shubik, M. (1972). "The assignment game I: The core." *International Journal of Game Theory* 1:111–130.

[139] Singer, I. (1997). *Abstract Convex Analysis.* Wiley.

[140] Sion, M. (1958). "On general minimax theorems." *Pacific Journal of Mathematics* 8:171–176.

[141] Strassen, V. (1965). "The existence of probability measures with given marginals." *Annals of Mathematical Statistics* 36:423–439.

[142] Tervio, M. (2008). "The difference that CEOs make: An assignment model approach." *American Economic Review* 98(3):642–668.

[143] Tinbergen, J. (1956). "On the theory of income distribution." *Weltwirtschaftliches Archiv* 77:155–173.

[144] Townsend, R. M. (1994). "Risk and insurance in village India." *Econometrica* 62:539–592.

[145] Trudinger, N. (2014). "On the local theory of prescribed Jacobian equations." *Discrete and Continuous Dynamical Systems Series A* 34(4):1663–1681.

[146] van der Vaart, A. (2000). *Asymptotic Statistics.* Cambridge University Press.

[147] Vershik, A. (2013). "Long history of the Monge–Kantorovich transportation problem." *Mathematical Intelligencer* 35(4):1–9.

[148] Villani, C. (2003). *Topics in Optimal Transportation.* Lecture Notes in Mathematics. American Mathematical Society.

[149] Villani, C. (2009). *Optimal Transport: Old and New.* Grundlehren der Mathematischen Wissenschaften, 338. Springer, Berlin.

[150] Vohra, R. (2005). *Advanced Mathematical Economics.* Routledge.

[151] Vohra, R. (2011). *Mechanism Design. A Linear Programming Approach.* Cambridge University Press.

[152] von Neumann, J. (1928). "Zur Theorie der Gesellschaftsspiele." *Mathematische Annalen* 100:295–320.

[153] von Neumann, J. (1953). "A certain zero-sum two-person game equivalent to the optimal assignment problem." In *Contributions to the Theory of Games,* H. W. Kunh and A. W. Tucker (eds.), Vol. II, pp. 5–12. Princeton University Press.

[154] Zambrini, J.-C. (1986). "Stochastic mechanics according to E. Schrödinger." *Phys. Rev. A* 33:1532–1532.

Index